ULTRASONIC TISSUE CHARACTERIZATION

Ultrasonic Tissue Characterization

Proceedings of the Second European Communities Workshop
30 November - 2 December 1981, Nijmegen, The Netherlands

Edited by

J.M. THIJSSEN and D. NICHOLAS

1982
MARTINUS NIJHOFF PUBLISHERS
THE HAGUE / BOSTON / LONDON

Distributors:

for the United States and Canada

Kluwer Boston, Inc.
190 Old Derby Street
Hingham, MA 02043
USA

for all other countries

Kluwer Academic Publishers Group
Distribution Center
P.O. Box 322
3300 AH Dordrecht
The Netherlands

Library of Congress Cataloging in Publication Data CIP

Main entry under title:

Ultrasonic tissue characterization.

Sponsored by the Committee on Medical and Public Health Research through the Specialized Working Group on Bioengineering, Evaluation of Technology, Transfer, and Standarization.
1. Diagnosis, Ultrasonic--Congresses.
I. Thijssen, J.M. II. Nicholas, D. III. E.C. Workshop on Ultrasonic Tissue Characterization (2nd : 1981 : Nijmegen, Netherlands) IV. Commission of the European Communities. Specialized Working Group on Bioengineering, Evaluation of Technology, Transfer, and Standardization.
RC78.7.U4U435 1982 616.07'543 82-14552
ISBN-13: 978-94-009-7668-9 e-ISBN-13: 978-94-009-7666-5
DOI: 10.1007/978-94-009-7666-5

FOREWORD

The second E.C. Workshop on Ultrasonic Tissue Characterization was organized for the assessment of research goals and plans as the basis to a grant proposal for a "Concerted Action" which will be forwarded to the European Commission. It should, therefore, not only be considered as a work in progress exhibition which succeeded the former activities within the E.C. (Inventory on Tissue Characterization in Europe and first Workshop).

The Steering Group invited a limited group of Institutes to send its group leader and a research fellow in the entitled field, who should expose the overall research program, as well as the more specific work on tissue characterization by RF signal analysis and spectroscopy. The ultimate goal of the Workshop has been the formulation of schemes of interaction among the Institutes and individuals and the definition of cooperative research programs. The remainder of the groups and some of the groups present are working in the field of analysis and processing of clinical A- and B-mode echograms (i.e. video signals). Because this work is more closely related to the clinical diagnostics it was decided that for that reason no preparatory workshop was needed. In the final proposal for the Concerted Action this work will get ample attention.

Although not many clinicians were present, their positive interest in the sophisticated analysis techniques and their contribution to the discussions again have been a most valuable and stimulating experience. We may emphasize that also in our possible future proceeding of the joint EC program the clinical interest and contribution will be stimulated and even be enhanced still further.

The organizer of the Workshop was backed and stimulated by the other members of the Steering Group, which at the time of the Workshop was composed as follows :

Prof. L. Masotti, Ph.D. (Florence)
D. Nicholas, Ph.D. (Sutton)
J.M. Thijssen, Ph.D. (Nijmegen), President
Prof. H.G. Trier, M.D. (Bonn)

The organizing committee worked enthousiastically and efficiently and I like to thank very much Miss Anja Derks, Miss Jolanda Hennink and Mr. René van den Broek, Marius Cloostermans and Wim Verhoef.
Last, but not least, Marion van Tongerloo, did all the typing and preparation work for the Proceedings Book. The editors and the authors are very much indebted to her for the high quality of the work she has delivered.

The Members of the Specialized Working Group on Bioengineering, Evaluation of Technology, Transfer and Standardization (SWG-BME), presided by Prof. D. Laurent (Créteil), are warmly thanked for their positive attention to our initiative and their effective channeling of it to adequate decision levels.

Nijmegen, May 1982 — J.M. Thijssen

CONTENTS

LIST OF PARTICIPANTS

Berger, G.
Laboratoire de Biophysique, ERA 498, CHU Cochin, Paris 14, France

Bredael, I.
European Communities Joint Research Center, Ispra, Italy

Castellini, G.
Istituto di Ricerca ulle Onde Elettromagnetiche, Consiglio Nazionale delle Richerche, Firenze, Italy

Clément, M.J.-M.
Laboratoire de Biophysique, ERA 198, Université de Paris-V, Paris, France

Cloostermans, M.J.T.M.
Biophysics Laboratory of the Department of Ophthalmology, University of Nijmegen, Nijmegen, The Netherlands

Cosgrove, D.O.
Institute of Cancer Research, Royal Marsden Hospital, Sutton, Surrey, United Kingdom

Decker, D.
Institut für Biomedizinische Technik, Stuttgart, FRG

Hill, C.R.
Institute of Cancer Research, Royal Marsden Hospital, Sutton, Surrey, United Kingdom

Irion, K.M.
Institut für Biomedizinische Technik, Stuttgart, FRG

Keuwez, J.
Université Libre de Bruxelles, Hôpital Universitaire Brugmann, Dept. de Gynaecologie-Obstetrie, Service d'Ultrasonographie, Fondation Freserh, B-1020 Bruxelles, Belgium

Landini, L.
Centro per l'Automatica 'E. Piaggio', Faculty of Engineering, C.N.R. Institute of Clinical Physiology at the University of Pisa and Institute of 1st Medical Pathology of the University of Pisa, Pisa, Italy

Lavelle, S.M.
Representative of the Specialized Working Group on Bioengineering, Evaluation of Technology Transfer and Standardization (SWG-BME), Dept. of Experimental Medicine, University College, Galway, Ireland

Lefébvre, J.-P.
Laboratoire de Mécanique et d'Acoustique, Marseille, France

Lepper, R.D.
Klinisches Institut für experimentelle Ophthalmologie der Universität Bonn, Bonn,FRG

Lerski, R.A.
Department of Clinical Physics and Ultrasonic Unit, Department of Diagnostic Radiology, Western Infirmary, Glasgow 611 6NT, United Kingdom

Lewin, P.A.
The Industrial Acoustics Research Group, Technical University of Denmark, Building 352, DK-2800 Lyngby, Denmark

Masotti, L.
Istituto di Elettronica, Facoltá di Ingegneria, Universitá degli Studi di Firenze, Firenze, Italy

Mesdag, P.R.
Delft University of Technology, Dept. of Applied Physics, Group of Acoustics, Delft, The Netherlands

Mol. J.
Representative of Ad Hoc Working Group on Doppler, Dept. of Clinical Neurophysiology, University of Limburg, Maastricht, The Netherlands

Morley, P.
Department of Clinical Physics and Ultrasonic Unit, Department of Diagnostic Radiology, Western Infirmary, Glasgow 611 6NT, United Kingdom

Nichols, D.
Institute of Cancer Research, Royal Marsden Hospital, Sutton, Surrey, United Kingdom

Picano, E.
Centro per l'Automatica 'E.Piaggio', Faculty of Engineering, C.N.R. Institute of Clinical Physiology at the University of Pisa and Institute of 1st Medical Pathology of the University of Pisa, Pisa, Italy

Rasmussen, S.N.
Rigshospitalet, University of Copenhagen, Dept. of Medical Gastroenterology, Blegdamsvej 9 DK-2100 Copenhagen, Denmark

Thijssen, J.M.
Biophysics Laboratory of the Department of Ophthalmology, University of Nijmegen, Nijmegen, The Netherlands

Trier, H.G.
Klinisches Institut für experimentelle Ophthalmologie der Universität Bonn, Bonn, FRG

Verhoef, W.M.
Biophysics Laboratory of the Department of Ophthalmology, University of Nijmegen, Nijmegen, The Netherlands

Van der Wal, L.
Delft University of Technology, Dept. of Applied Physics, Group of Acoustics, Delft, The Netherlands

Concise Report

2nd Workshop on Ultrasonic Tissue Characterization
30 November - 2 December 1981

Sponsored by the Committee on Medical and Public Health Research (CRM) through the specialized Working Group on Bioengineering, Evaluation of Technology, Transfer and Standardization (SWG-BME).

This report has been written by J. M. Thijssen, Ph.D., chairman of the Planning Group and the local organizing committee.

The scope of the Workshop

Ultrasonographic tissue characterization has evolved rapidly since our first effort to reveal the state of the art within the European Communities two years ago.
In the clinical diagnostic field the growing interest in calibration and standardization facilities signals the awareness clinicians have of tissue signature in the high quality gray scale B-mode pictures. The manufacturers are responding similarly by implementing software techniques currently utilized in other diagnostic imaging systems, such as CT-scanning and emission tomography. They not only provide the clinician with various options for postprocessing for the video images, but also analysing algorithms such as regression and texture statistics. It may be noted, therefore, that one of the conclusions of the first workshop is being aided by the great potential of modern echographic equipment, i.e. the expansion of the facilities of clinical equipment towards better images and more quantitative information. Analysis of the radiofrequency echograms has evolved to in vivo (clinical) applications in many ways. Diffraction measurements in angular real time scanning is most promising and it may yield a valuable structural parameter in tissue characterization. Attenuation

coefficient estimation from backscattered ultrasound is presently being implemented in clinical trial already in some institutions. Impediography and interface/membrane characterization will become useful clinical tools and may become important in the differentiation of focal lesions. Analysis of tissue motility has attracted interest and may yield a worthwhile differentiating parameter as well. In vitro experiments with healthy and diseased (tumour) tissues are performed in several research centres and allow the practical limits of the various methods to be estimated and their clinical feasibility assessed.
Furthermore, theoretical and experimental work on reflection and transmission reconstructive techniques are beginning to produce results. In both the synthetic aperture focussing and the computerized tomography progress is greatly supported by the rapid developments in microelectronics and computer hardware.

The 12 participating institutions have been selected from a total number of 21 institutions who responded positively to an invitation to become involved in an eventual concerted action. The selection was made in such a way that a homogeneous group was formed of people working in a clinical environment on RF echogram analysis. This broad subject asks for a long duration action of coordination and cooperation within the E.C. The work is in various stages of in vitro feasibility studies, of theoretical studies on the interaction of ultrasound with tissues and even of clinical pilot studies.
It may be remarked that the planning of the Concerted Action should be molded in such a way that short time (at least in technological sense) activities are involved and clinical trials should be coordinated and expanded.

The lectures

The schedule of the presentations is given in the abstracts booklet. The lecturers gave a global survey of the ongoing work in their institution which was complemented by a detailed description of the tissue characterization activities. Presented were details concerned with the equipment facilities, the physics and mathematics of the computer analysis techniques and the results of in vitro experiments and of in vivo clinical pilot studies. Finally the planning of future activities was outlined. The discussions on the lectures took approximately one third of the total duration of the workshop which is indicative of the fruitful interactions among the participants.

The subjects presented in the lectures comprised in general various techniques for processing of the radiofrequency echograms. Some aspects also involved the devisions of transmission and reflection reconstruction techniques aiming at either improved echographic imaging of tissue inhomogeneities, or imaging of impedance, velocity, and attenuation profiles (ultrasonic CT scanning). These techniques are to be encouraged since they provide an extension into the second dimension of the scanning plane of the single trace RF analysis as presented by other groups.

Other techniques discussed concern the characterization of membranes and interfaces by means of deconvolution, correlation and spectral analysis. The application in ophthalmology is already obvious, but it may be conceived that the investigation of encapsulated small focal lesions can be improved by these techniques also.

The instrumentation needed for future clinical application of the various techniques was discussed, and it appeared that some groups have a great potential and know-how to proceed in the development of dedicated equipment.

The problem of statistical methods to remove part of the effects of the tissue inhomogeneity on the one hand and the extension to second order statistics to enhance the reliability of texture analysis on the other hand received ample attention. It is to be expected that the sophistication of statistical analysis and decision procedures will play a major part in the long run in computer aided ultrasonic tissue characterization.

Conclusions from the General Discussion

The discussions were devoted to instrumentation and techniques, and to two main fields of research which, for convenience, were termed pattern recognition applied to echography and the estimation of acoustical tissue signature parameters.

This division is somewhat arbitrary but allows the different themes of the topic to be discussed in detail. Obviously there is a degree of overlap between these two fields, this is to be explained by the presence of several representatives of groups active in both fields. On the other hand it should be stressed that the list of groups <u>potentially</u> involved in a concerted action scheme should be reconsidered and a more complete list was produced.

The final proposal for a concerted action will be based on two <u>topic oriented research groups</u>, entitled :

1. Pattern recognition of echograms
2. Acoustical tissue signature parameters

The activities within these groups will proceed along the following general line :

- development of measurement and analysis techniques
- in vitro experiments + evaluation
- in vivo pilot study + evaluation
- prolonged clinical study (prospective, correlation)

All of these steps will be carried out in the form of multicenter coordinated activities.
The above scheme is applicable to subactivities either on a single tissue/organ in case of pattern recognition and to a single tissue parameter in the work in the other research group. It is possible, however, that some of these activities will run in parallel.

A lot of ideas arose from the discussions which concern: exchange of data (file structure), computer programs, tissue phantoms, calibration and standardization etc. It is considered of high relevance to incorporate exchange visits to be able to perform measurements with "take round specimens", and to check the equivalence of equipment. Basic requirements for the concerted action to become fruitful are exchange visits, experts meetings and (regular) workshops of the (sub)groups.

The presence of three representatives (Prof. Mol of the Ad-hoc Working Group on Doppler, Prof. Lavelle of SWG-BME, and Mr. Bredael from ISPRA) was highly appreciated. The participants decided that it would be very important for future work in the E.C. if the non-destructive testing group from ISPRA could become involved. Especially in the field of equipment testing (standards, calibration) their contribution might become of great value.

TRANSMISSION AND REFLECTION TECHNIQUES FOR ATTENUATION MEASUREMENTS IN TISSUE CHARACTERIZATION

G. Berger, J. Perrin and D. Bourgoin
Laboratoire de Biophysique, ERA 498, CHU COCHIN, Paris 14e, France

Introduction

The first results (1,2,3) we obtained for experimental tissue perturbations showed the role played by cellular organization in the attenuation coefficient of soft tissues, encouraging us to performe an additional series of "in vitro" experiments.
Simultaneously, we built a device for acquiring classical A scans, which acquires, at a high frequency rate, the reflected signals.
Finally we tested different statistical methods leading to the evaluation of the slope of attenuation versus frequency for the collected signals.

TRANSMISSION ATTENUATION MEASUREMENTS- "IN VITRO" TISSUE PERTURBATIONS.

1. Methodology

The equipment used for transmission measurements is briefly described (Fig. 1).

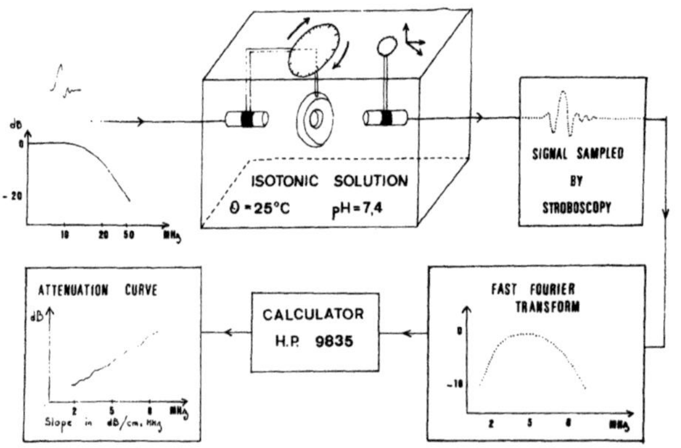

Figure 1. Equipment for "in vitro experiments".

The emission pulse delivered by a Metrotek module, has a bandwidth of 11 MHz at -3 dB. The transmitted signal, sampled by stroboscopy, is Fast Fourier Transformed and the resulting spectrum has a central frequency of 5 MHz and a frequency bandwidth of 6 MHz at -6 dB. The attenuation is obtained by a classical spectroscopic method, where a linear regression of the data, operated in the range of frequency where the standard deviation is small, gives the attenuation slope in dB/cm.MHz.
The biological experiments, performed on a multilobulated rabbit liver allowed us to obtain a reference sample from one lobe, then to inject a perturbing solution through the vascular system of the other lobes and to extract the perturbed sample for comparison. The aim of our pertubations has been to modify the intercellular boundaries of the liver.
All our samples were histologically observed to confirm the efficiency of the cellular dissociation without any extraneous effects on the cells.

2. Experimental results

a) Trypsine injection (0.5%)

A solution of trypsine, whose effect is known to break the chemical cell boundaries, has been injected. After histological checks, the attenuation slopes were compared for both reference and perturbed sample (Table 1). All these experiments exhibited a significant drop of the attenuation coefficient which will be discussed later on.

Table 1 . Comparison of the attenuation slope of the reference sample (A_r) and the sample with reduced intracellular junctions (A_p)

EXPERIMENT	Limits (F_1, F_2) MHz	Attenuation slope REFERENCE sample (dB/cm.MHz) A_r	Attenuation slope PERTURBED sample (dB/cm.MHz) A_p	$\frac{A_p - A_r}{A_r}$
n°1	3.5-6	1.71	1.13	-0.34
n°2	4-7	1.12	0.50	-0.55
n°3	3-7	1.31	0.72	-0.45
n°4	3-8	1.10	0.77	-0.31
n°5	3-7	1.40	1.01	-0.28

b) Action of collagenase-hyaluronidase and chelators of calcium

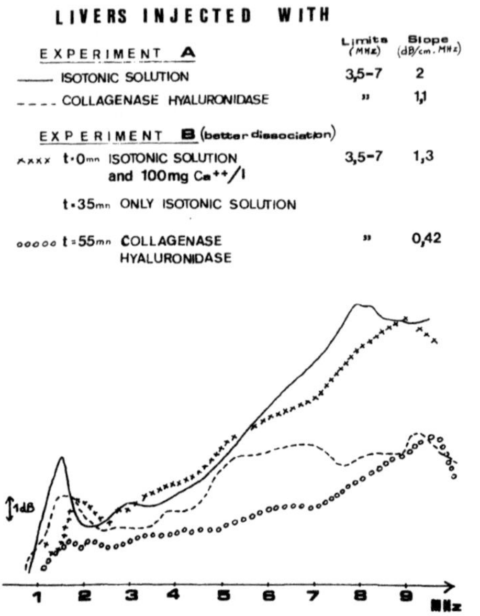

Figure 2. Successive destruction of the chemical and electrostatic forces.

Let us recall that intercellular junctions are mainly composed of chemical and electrostatic forces, on which we have successively acted.

In the experiment A (Fig. 2) the chemical forces are destroyed by the action of collagenase-hyaluronidase, and in the experiment B, we added the action of chelators of calcium to destroy the electrostatic ones. This biological procedure has been described by Seglen (4,5).

c) Results and discussion

From our experimental work the following points can be established :

1. The attenuation slope for normal liver (Table 1) varies from 1.1 to 1.7 dB/cm.MHz which is most certainly correlated to a natural variation from one liver to the other.
2. The percentage drop of the attenuation between normal and dissociated tissue varies by 40% ± 12%. That means that even the smallest value (28%) is still a significant drop, presumably due to the applied cellular dissociation. The proportion of dissociated cells varies in the various experiments and may explain the variations of the percentages.
3. When we experimentically increased the cellular dissociation (Fig. 2), the drop in attenuation increased. This seems to imply a quantitative relationship between the slope of the attenuation and the degree of cellular dissociation.

4. To confirm this fact, we followed in time the evolution of the attenuation slope of the same sample while the enzyme was carrying out its action (Fig. 3). We obtained a pronounced drop during the first 20 min., followed by a rise which still has to be explained.

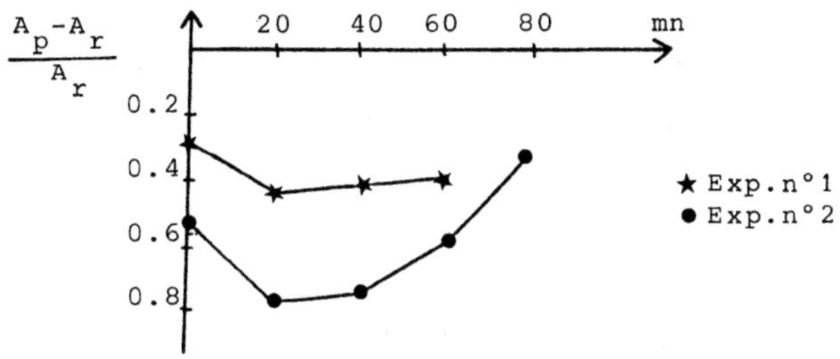

Figure 3. Evolution in time of attenuation slope of perturbed tissue relative to that of non-treated reference tissue, both measured at the same time after excision.

Finally we can conclude that attenuation is operating at the cellular level, probably through cellular relaxation mechanisms. Cancer is an example of pathology where the cellular cohesion is modified, as cells are able to move through the human body. But this tissue modification, which reduces attenuation, is complemented by a structural modification (neovascularisation) which increases attenuation through scattering mechanisms. Thus, in this case, only scattering and attenuation may provide useful information.

SCATTERING-ATTENUATION MEASUREMENTS

First we built a digitizer working at 24 MHz which was incorporated into a classical B scanner. Then we tested methods for performing attenuation measurements on reflected signals.

1. Signal acquisition

A region of interest is selected on a B mode echogram by positioning a marker in the image (Fig. 4). During a subsequent A-scan, a region of 8 mm of tissue is converted and transfered to a digital computer and a Fast Fourier processor.

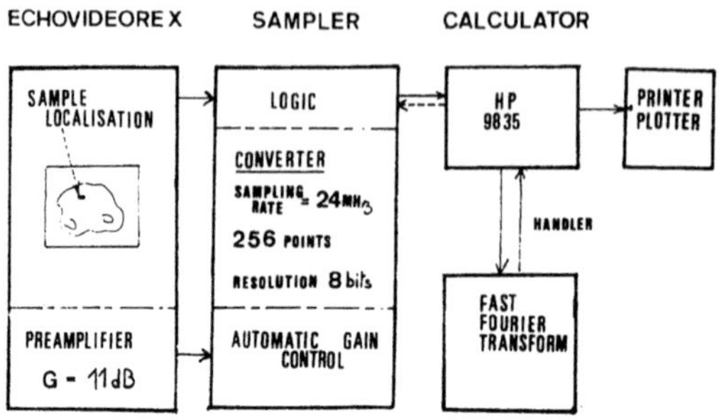

Figure 4. Equipment for "in vivo" measurements

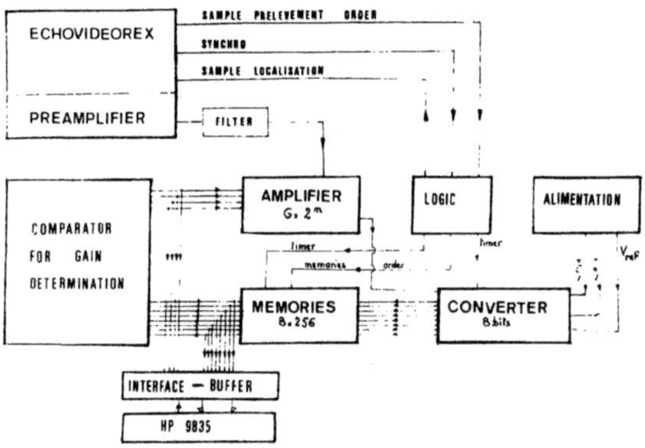

Figure 5. Details of the sampler

The details of the digitizer are presented in Fig. 5. The conversion rate is 24 MHz and it stores 256 samples of 8 bits. A filter with 3 dB fall off at 11 MHz and 70 dB at 13 MHz has been introduced between the pre-amplifier and the digitizer.

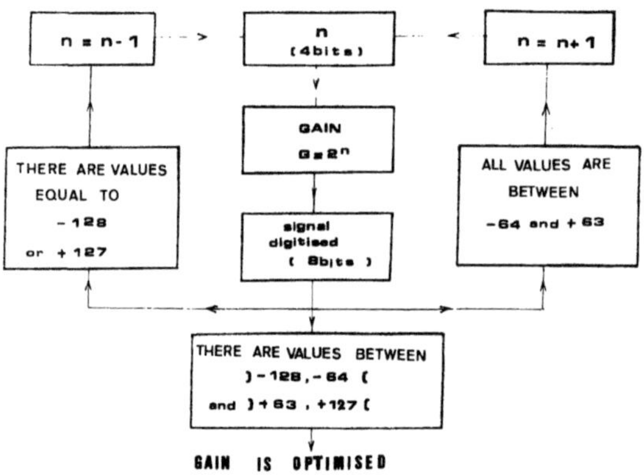

Figure 6. Automatic gain control

In order to make the best use of the dynamic range of the converter the sample is submitted to amplification throughout its 8 mm length, determined by an automatic gain control. This amplification is controlled by a 4 bits value "n", the optimisation of which is explained in Fig. 6. The overall dynamic range is 78 dB.

2. Methods and experiments

a) Correlation reflection - transmission

We studied on the same sample the transmitted signals and the signals reflected from the interfaces and internal reflectors of the structure, simultaneously. Deconvolution between interfaces echoes gave an attenuation slope in good agreement with the transmission measurements but very different from the values obtained from using internal echoes. This experiment clearly shows the role played by the reflection transfer function as a function of frequency.

b) Angular averaging method developed from specular interfaces

We will explain the transfer function between two specular interfaces which we can consider as presenting a reflection angle θ with the incident beam. Such reflectors could represent interfaces between two organs, vessels or primarilly collagenic structures.

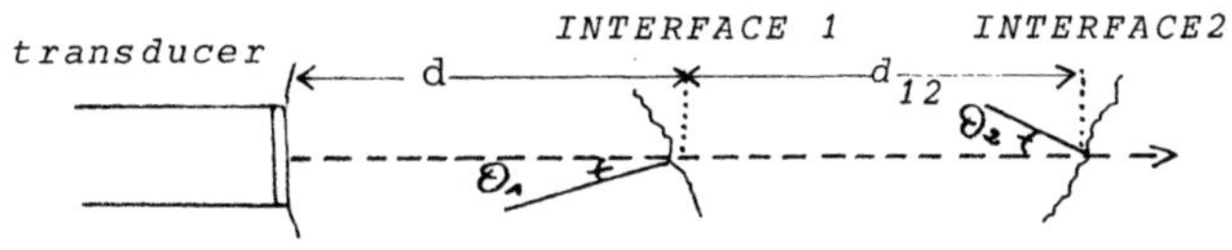

Figure 7. General scheme of reflection with two interfaces

If we define (Fig. 7) :

- . $H_{12}(f)$ as the attenuation function over the distance d_{12}
- . $R_{d,\theta}(f)$ as the reflection function at the interface (which is dependent on θ and also on d in the near field of the transducer)
- . $Y_1(f)$ and $Y_2(f)$ as the spectra of the signals received from the interfaces 1 and 2, respectively.

Then we obtain :

$$\frac{Y_2(f)}{Y_1(f)} = \frac{R_{\theta_2}(f)}{R_{\theta_1}(f)} \cdot H^{*}_{12}(f) \cdot H_{12}(f)$$

$R_{\theta_2}(f)/R_{\theta_1}(f)$ is perturbing the interpretation of the reflection especially because of its dependence upon frequency, which is described below.

The interfaces are modelled as rough surfaces characterized by a roughness "h". The theoretical study of such surfaces (6,7,8,9,10) has led to the expression of the reflected intensity for an angle θ (I) compared to the incident intensity (I_0) :

$$\frac{I}{I_o} = \frac{1}{\cos\theta\left[1 + t_g^2\,\theta\,(K/R^2 f^{2n})\right]} = |R_\theta(f)|^2$$

n and K being functions of the surface.

The theoretical curves have been plotted (Fig. 8 and 9) and validated by experimental angular reflection diagrams. These diagrams obtained with liver from -10° to + 10° are in agreement with the model presented here. For this case, if we compare the angular averaged spectra they give :

$$\frac{\overline{Y}_2(f)}{\overline{Y}_1(f)} = \frac{\frac{1}{N_2}\sum_{\theta_2} R_{\theta_2}(f)}{\frac{1}{N_1}\sum_{\theta_1} R_{\theta_1}(f)} \times H_{12}^2(f)$$

The averaged reflection functions have been simulated (Fig. 10), where we can conclude that :

- no significant difference appears after 10° of summation
- the variation with frequency of the averaged function is smoother
- if statistical parameters of both interfaces are similar, the averaged reflection functions have the same dependence on frequency and we have that :

$$\frac{\overline{Y}_2(f)}{\overline{Y}_1(f)} = K' \times H_{12}^2(f)$$

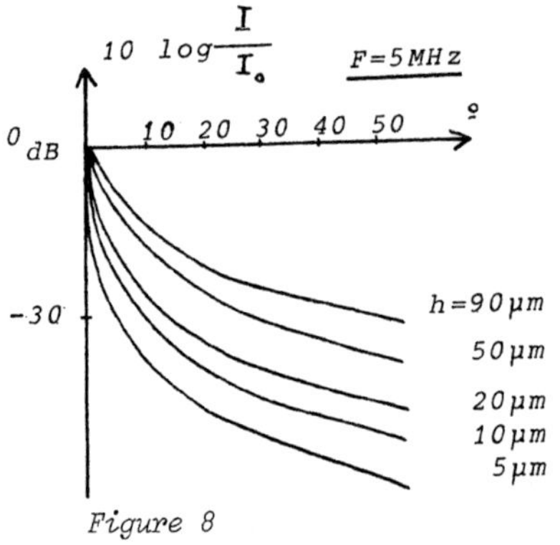

Figure 8

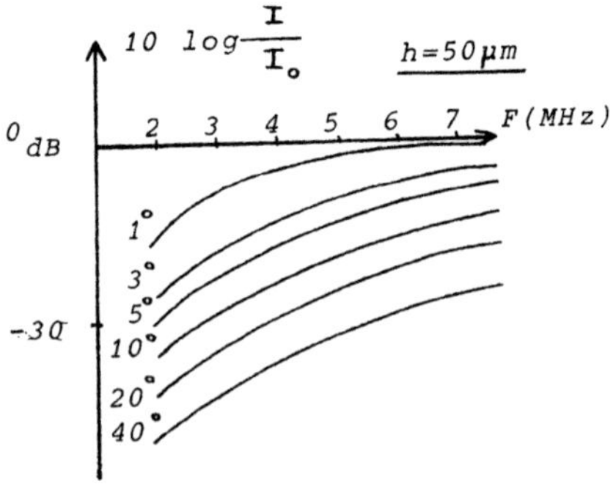

Figure 9

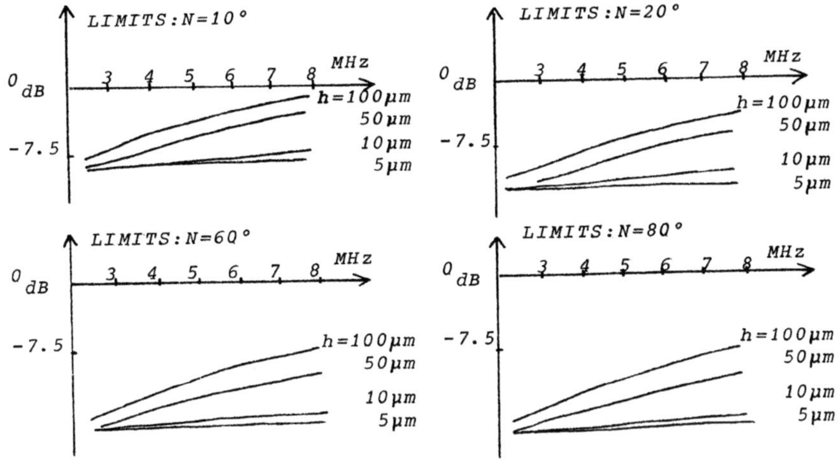

Figure 10. $10 \log \left| \frac{1}{N} \Sigma \frac{I}{I_0} (\theta) \right|$ *versus frequency at various values of h for N = 10,20,60,80*

This method has been applied to phantoms constructed from liquid filled fingers. From our measurements we have extracted the attenuation slopes of different solutions. Fig. 11 presents the experiments performed with two macromolecular solutions at different concentrations.
We also performed this experiment on gloves filled with water, to prove that no artefacts were introduced by the method.

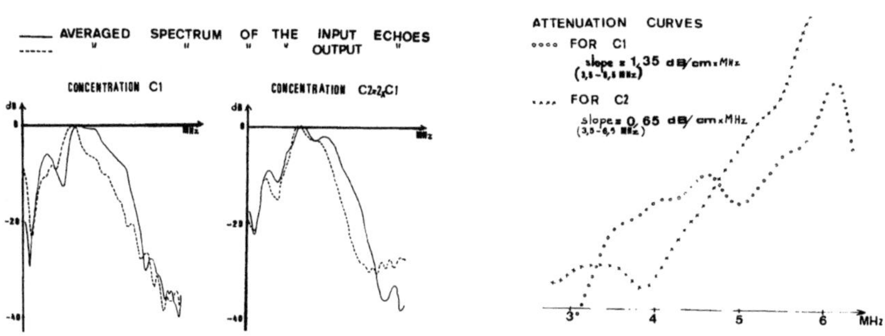

Figure 11. Averaged spectra and resulting attenuation curves

c) Best estimates of attenuation slope

If we receive N echoes from a scattering medium, the relation, using the precedent notations, between any two spectra (i,j) is still given by :

$$\frac{Y_i(f)}{Y_j(f)} = \frac{R_i(f)}{R_j(f)} \times H_{ij}^2(f)$$

The logarithmic expression becomes :

$$10\log Y_i^2(f) - 10\log Y_j^2(f) = 10\log R_i^2(f) - 10\log R_j^2(f) + 2d_{ij}A_{ij}(f)$$

where $A_{ij}(f)$ is the attenuation in dB/cm measured at the distance d_{ij}.

Kuc and Schwartz (11,12) made a theoretical study of $10\log R_i^2(f) - 10\log R_j^2(f)$ which is treated as a white Gaussian process.

So each pair of values (i,j), the number of which varies from 1 to K, provides an atttenuation curve $A_k(f)$ corresponding to a distance "d_k" and yields a slope $\hat{A}_k$ in dB/cm.MHz.

The theoretical expression for the best estimates of slope $\acute{A}$ and its variance $\mathrm{var}(\tilde{A})$ are given by the expressions :

$$\tilde{A} = \frac{\sum_{k=1}^{K} d_k{}^2 \hat{A}_k}{\sum_{k=1,k} d_k{}^2}$$

$$\dot{\mathrm{var}}(\tilde{A}) = \frac{K'}{\sum_{k=1}^{K} (2d_k)^2} = \sigma^2$$

where K' is a constant for a particular tissue and for fixed regression limits.

We have experimented with this procedure on freshly excised porcine livers. The fact that the organ is excised allows us to work outside of the diffraction effect zone and the available size of porcine liver allows us to use a great number of measurements.

The best estimate of slope (A) and its experimental standard deviation (σ) are calculated for each organ (Table 2). If we consider the variance formula, it appears that a great number of differences minimizes σ but in this case, some of them correspond to small distances d_k which increase σ. We have approached this optimum by taking into account a selection of these differences (Table 2).

Table 2. Results obtained on porcine liver

PORC LIVER	A ALL the differences are treated			B SELECTION of the differences according to the distance		
	K	dB/cm.MHz		K	dB/cm.MHz	
		$\tilde{A}$	σ		$\tilde{A}$	σ
n°1	40	0.82	0.16	25	1.00	0.13
n°2	30	1.08	0.22	22	0.88	0.17
n°3	35	0.79	0.15	23	0.69	0.16
n°4	34	0.60	0.11	26	0.56	0.10
Mean value		0.82			0.78	
Dispersion		±0.20			±0.22	

From Table 2, the following points can be established :

1) Variations between experiments (0.2 dB/cm.MHz) may be attributed either to natural differences (cf. in vitro results) or to the methodology (we have the same variations for each experiment).
2) The suppression of small distances for the best estimates calculation does not affect the mean value (0.82 to 0.78) but modifies the value of each experiment.
3) The value of σ is smaller in the case B, but still too large to permit a fine tissue characterization.

d) Short Time Fourier Analysis and Center of Gravity method

Three inconveniencies associated with the preceding procedures must be presented :

1) The role played by the time window for echo selection is not at all neglegible, especially when two echoes are included yieling a bimodal spectrum.
2) The slope is very sensitive to any spectral distortions.
3) The diffraction effect, which modifies the spectra according to the depth of the tissue traversed, even when no attenuation exists, can not be taken into account.

We are now going to introduce a new method developed by Hottier et al. (13) in the L.E.P. laboratory (France). This method is based on the evolution with depth of tissue of the center of gravity of Short Time Spectra (Fig. 12).

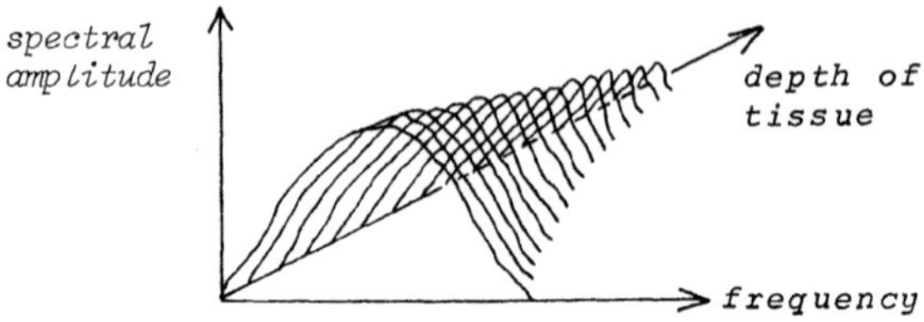

Figure 12. Schematic evolution of spectra with the depth of tissue

Here the total signal is explored, giving the following advantages :

1) The amount of information per unit distance is larger, which allows us to interrogate SMALLER SLICES of tissue.
2) The continuous information may show finer attenuation DISCONTINUITIES.

The evolution of the center of gravity (C.G.) with the depth of tissue, exhibits (Fig. 13a) a decrease directly correlated to the slope of attenuation.

1) This parameter minimizes the influences of spectral distortions and time window effects from echo selection.
2) Taking the diffraction effect into account we remark (Fig. 13b) that the evolution of the C.G. is modified, especially in the near field. Nevertheless, it is possible by this method to deconvolve the effect of diffraction and obtain the linear dependence again.

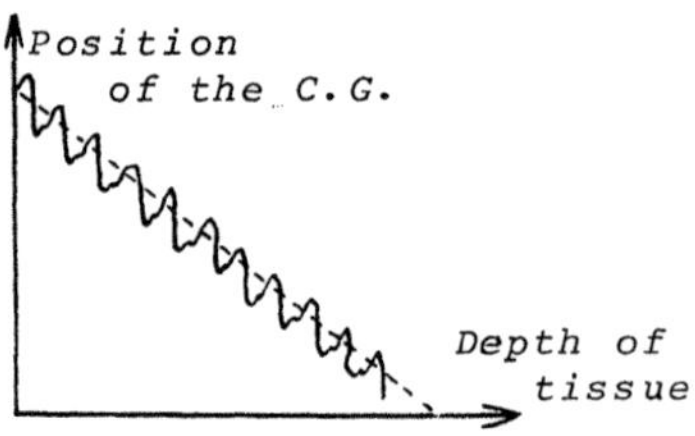

Figure 13a. Schematic evolution of the C.G. without diffraction

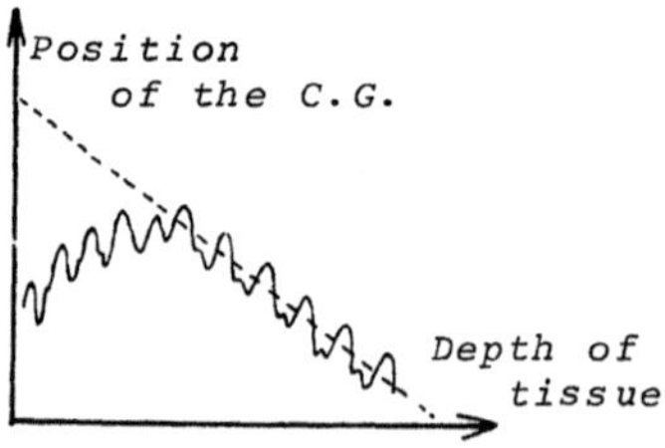

Figure 13b. Schematic evolution of the C.G. with diffraction

The first experiments, performed by Fink on phantoms are encouraging and we are soon going to collaborate for "in vivo" acquisitions.

CONCLUSION

In the first part of this paper we have shown the role played by cellular organization, especially through the intercellular forces on attenuation measurements. The actual performance of "in vitro" work is based on a systematic study of rat liver cancers at different stages.
In the second part, we have then presented a device, integrated to a B-scanner, for reflected signal acquisition. We have shown that angular averaging could minimize the effect of the reflection transfer function on the frequency dependence. We tested, avoiding diffraction effects, a method developed by Kuc to find the best estimates of the attenuation slope, but we experimentally obtained large standard deviations. Finally we introduced a method based on Short Time Fourier Analysis which monitored the evolution with depth of tissue of the gravity center of the spectra. The advantages of the latter method that we have presented led to the preliminary conclusion that valuable clinical applications may arise.

REFERENCES

1) Perrin, J., Berger, G., Salesses, A. and Agneray, M. (1979) : Role played by coherent and incoherent diffusion in ultrasonic attenuation of biological tissues. Ultrasonics International 79 - Conference proceedings (IPC Science & Technology Press, Guilford), 518-523.
2) Berger, G. and Perrin, J. (1979) : Attenuation principles and measurements for tissue characterization. Ultrasonic Tissue Characterization, (Ed. J.M. Thijssen. Stafleu, Alphen a/d Rijn/Brussels), 177-185.
3) Berger, G., Perrin, J., Bourgoin, D., Salesses, A., Agneray, A. and Darlas, Y. (1982): Role played by cellular cohesion and organisation in the interaction of ultrasound with soft tissues. Acoustical Imaging, Vol 10 (Ed. A. Metherell. Plenum Press, New York), 253-264.
4) Seglen, P.O. (1972) : Preparation of rat liver cells. Experimental Cell Research, 450-454.

5) Seglen, P.O. (1973): Preparation of rat liver cells 2. Effect of ions and chelators on tissue dispersion. Experimental Cell Research 76, 25-30.

6) Quentin, G., de Billy, M., Cohen-Tenoudji, F., Doucet, J. and Jungman, A. (1975): Experimental results on the scattering of ultrasound by randomly and periodically rough surfaces in the frequency range 2 to 25 MHz. Ultrasonics Symposium Proceedings, IEEE, 102-106.

7) de Billy, M., Cohen-Tenoudji, F., Jungman, A. and Quentin, G. (1976): The possibility of assigning a signature to rough surfaces using ultrasonic backscattering diagrams. IEEE, Transactions on Sonics and Ultrasonics, Vol. SU-23, 356-363.

8) Eaglesfield, C.C. (1962): The angular dependence of scattering by nearly polished surfaces, I. Proceedings Physical Society, Vol. 80, 947-951.

9) Eaglesfield, C.C. (1963): The angular dependence of scattering by nearly polished surfaces, II. Proceedings Physical Society, Vol. 82, 406-411.

10) Senapati, N., Lele, P.P. and Woodin, A. (1972): A study of the scattering of submillimetre ultrasound from tissues and organs. Proc. IEEE Ultrasonics Symp. Cat. Nr. 72 CHO 78-8SU, 59-63.

11) Kuc, R. and Schwartz, M. (1979): Estimating the acoustic attenuation coefficient slope for liver from reflected ultrasound signals. IEEE, Transact. on Sonics and Ultrasonics, Vol. SU 26, 353-361.

12) Kuc, R. (1980): Clinical application of an ultrasound attenuation coefficient estimation technique for liver pathology characterization. IEEE Transact. on Biological Engineering, Vol. BME-27, 312-319.

13) Hottier, F., Fink, M. and Berger, G.: Experimental estimation of attenuation in liver using short-time Fourier analysis : Influence of system parameters. Seventh International Symposium on Ultrasonic Imaging and Tissue Characterization, Gaithersburg, 1982. (Abstract to appear in: Ultrason. Imag.)

Discussion

G. BERGER, J. PERRIN and D. BOURGOIN

Ultrasonic transmission and reflection techniques for tissue characterization with B scan and for tomographic imaging

Question asked by: L.F. v.d. Wal
Your time gain amplifier covers a range of 90 dB. Could you tell me once more how you get to this number, using only 4 bits.

Reply by: G. Berger
With 4 bits we can generate 16 different gain settings. Each gain setting allows for a change of ca. 6 dB in the total gain. In this way we can achieve in theory gain settings ranging from 0 - 90 dB.

Question asked by: J.M. Thijssen
How is the AGC effectively working? Does it compare each sample to a wanted size, and does it work in real time during analog to digital conversion?

Reply by: G. Berger
The AGC is given a fixed value of the gain over samples of 8 mm length. Over this sample a variation gain of 6 x n dB is compensated after n iterations.
So, for each sample of 8 mm length, the signal is coded into 8 bits and the gain known into 4 bits.

Question by: C.R. Hill
Does your digital gain control reconstruct a scattering coefficient independent of the attenuation coefficient?

Reply by: G. Berger
No, the AGC is just a way to make the best use of the dynamic range of the converter.

Further discussions and remarks

The experiments on binding forces between epithelial cells of rats may perhaps be performed while using diagnostic frequencies (> 1 MHz) and (peak) intensities. The outcome may then become very important in the understanding of possible traumatic effects of ultrasound energy in biological tissues.

The technique of short time spectral analysis (S.T.S.A.) seems to produce statistically more reliable estimates of the attenuation coefficient frequency slope than the method of Kuc & Schwartz. Since STSA is to be considered to be a kind of smoothing procedure it remains to be proven whether this improvement is just apparent, or real.

A more general question on the comparability of the attentuation measurements with phase sensitive and phase insensitive receiving transducers appears still to be unsolved.

ULTRASONIC QUANTITATIVE IMAGING WITH A RING ARRAY FOR BREAST EXAMINATION

M. J.-M. Clément and J. Perrin
Laboratoire de Biophysique, ERA 198, Université Paris-V, Paris, France

Introduction

An alternative approach to echographic breast imaging can be found in computerized transaxial tomography which can provide quantitative cross sectional images for various bulk acoustic parameters pertaining to the tissue. The first ultrasonic tomograms have been recorded in-vitro by Greenleaf et al. (1) and laterly in-vivo (2). In order to eliminate mechanical scanning effects in tomographic data recordings, an immobile 360° ring array has been proposed (3,4). The system architecture for electronic steering and scanning such an array has been described adopting "constant-K" mode of scanning (5). In this paper, we report the manufacturing technique for a 450 element toroidal ring array which has been built, as well as its experimental acoustic properties. We will also compare the ray displacement effects, due to refraction at a breast interface, for both a ring and linear array.

Ring array transmission imaging

In X-ray and emission tomography, ray paths are straight lines but this is not usually so in ultrasonic tomography. The problem of ray curvature has been discussed by several researchers and various schemes have been reported to improve on this. In particular, it has been shown that there are "forbidden" regions into which "shortest" rays never penetrate so that it is impossible to reconstruct exactly the refractive index in these regions. As these refraction problems are not only dependent on object size and refractive index, but also on transducer separation, one might expect that a ring array geometry will be more favorable for circular objects such as breasts. This turns out to be the case when one considers the ray path displacement due to refraction at a breast interface (6).

Ring array design and construction.

The main design parameters of a ring array are identical to those for a linear one, with the addition of ring diameter and number of sectors the array is made of. The choice of the ring diameter is a result of conflicting constraints. For refraction considerations, the ring diameter must be as small as possible. On the other hand, to minimize the offset angle between the projection path and the array geometry, occuring at increasing values of the projection index K, the diameter must be as large as possible. We have retained a ring diameter of 220 mm.

The array we chose has been constructed from 5 independent and self contained 72° toroidal sectors of 90 elements. The manufacturing technique for these sectors has been described elsewhere (5). Since our first reported implementation and before the realization of the full array reported here, we have improved our soldering technique and thickness control of the front face of the array. The sectors are mounted onto a supporting plate which is then fitted into the immersion tank. During plate mounting, accurate control of the lateral alignment of the sectors must be ensured for a regular spacing of each array element along the whole circumference (6). The main characteristics of the final array are as follows : fo = 3.5 MHz, Ø = 220 mm, slice focussing radius of curvature = 110 mm, number of elements = 450, element height = 18 mm, element width = 0.53 mm. The array built is made of 5 sectors as shown in Fig. 1.

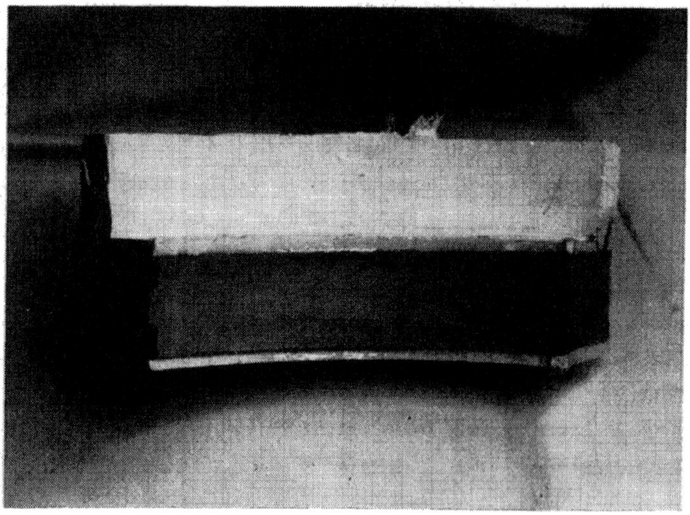

Figure 1 . A toroidal sector for ring array construction

Experimental system.

The system architecture of the electronics built to multiplex the constructed array has been described in a previous article (5). The main feature of this system is that several contiguous elements can be multiplexed and steered while a single element is selected as a receiver.

Experimental results and discussion.

As the ultrasonic tomographic technique encounters refraction, diffraction and reflection effects, there are implementation artefacts within the reconstructed image which are then dependent on the insonifying beam shape and characteristics, as well as the transducer separation (5). The main beam characteristics of the array are described experimentally in the following acoustic measurements. These measurements have been performed both with a single transmitter and 7 transmitters, with and without electronic steering. The ultrasonic amplitude profile across the tomographic slice, for distances ranging from 20 to 200 mm shows a maximum amplitude at 100 mm, with a 3 dB depth of focus of 110 mm ranging from 50 to 160 mm. The narrowest tomographic slice thickness is 4.2 mm and occurs at 110 mm. The 6 dB slice thickness at 20 and 200 mm is respectively 11.5 and 10 mm. Figures 2 and 3 have both been recorded using a single element along the ring array as the receiver and respectively in A, B and C, 7 unsteered transmitters, 7 electronically steered transmitters, 1 single transmitter along the ring. These diagrams are plotted as functions of the offset angle θ, or, equivalently, the projection index K (the offset angle θ being the angle between the projection path and the axis normal to the transmitter). Figure 2 is a plot of the amplitude A_o of the received signal versus θ, with the array immersed in water without an object. These experimental curves illustrate the gain in transmitted power when increasing the number of transmitters (20 dB between C and A-B). Moreover, it shows an unsteered 20 dB offset angular range of $\pm 7^o$ (A) (corresponding to a free water object field diameter of 27 mm), against a 20 dB offset angular range of + 38° for a steered array (B) (corresponding to a free water object field diameter of 172 mm). For time-of-flight tomography, the S/N ratio and the frequency of the received signal determine the time measurement accuracy. We have plotted in Figure 3 the frequency of the largest cycle of the received impulse versus K and θ. If the frequency variations are approximately doubled with several transmitters compared to one (with variations 3-4% smaller for steered (B) than unsteered transmitters (A)), the measurement accuracy will not be impared by using several transmitters instead of one, as within $\pm$ 40° θ spans, the frequency corresponding to several transmitters never gets smaller than for one.

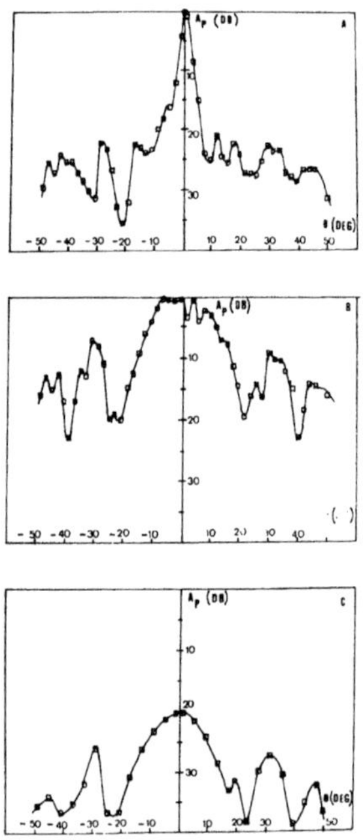

Figure 2. Signal amplitude versus offset angle

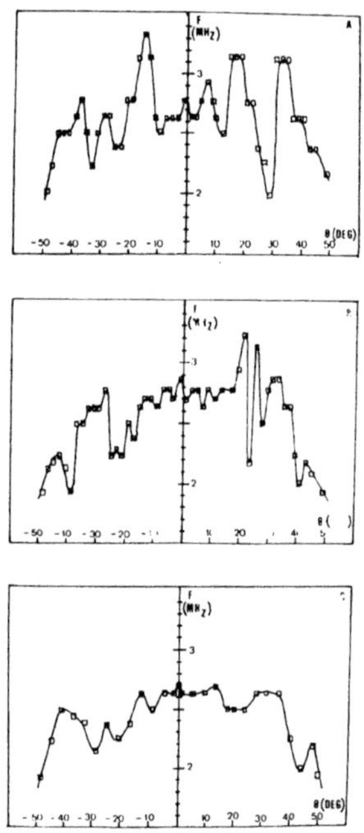

Figure 3. Signal frequency versus offset angle

CONCLUSIONS

The realization of a 3.5 MHz, 220 mm diameter, 450 elements, 360° toroidal ring array for ultrasonic transaxial tomography has been reported. Its refraction properties regarding circular objects have been compared to a linear array. It has been experimentally shown that electronic steering increases the 20 dB angular span of the array with 7 transmitters from $\pm$ 7° to $\pm$ 38° (corresponding to object diameters of 27 and 172 mm). The technique of ring array tomographic recording should prove to be very useful in the implementation of a new generation tomographic breast scanners, both from the imaging as well as from the tissue characterization point of view.

ACKNOWLEDGEMENTS

M.J.-M Clément would like to express his gratitude to Prof. P. Alais and Prof. J-C. Roucayrol for their encouragement and support. He is also very grateful for the technical assistance of Mr. Fruchard in mould manufacturing.

REFERENCES

1) Greenleaf, J.F., Johnson, S.A., Samayoa, W.F. and Duck, F.A. (1975): Alg. reconstruction of spatial distributions of acoustic velocities in tissues from their time-of-flight profiles. In: Acoustical Imaging, Vol. 6, 71-90. Ed. N. Booth, Plenum Press, New York.
2) Greenleaf, J.F., Kenue, S.K., Rajagopalan, B. et al. (1979): Breast imaging by ultrasonic computer assisted tomography. In: Acoustical Imaging, Vol.8, 599-614. Ed. A. Metherell, Plenum Press, New York.
3) Clément, M.J-M; and Alais, P. (1978): French Patent nr. 7818424, June 1978.
4) Clément, M.J-M., Alais, P. and Perrin, J. (1979): Ultrasonic computed tomography by electronic scanning of an annular array. Proc. Ultrasonics Int. 79, 511-517 (Graz).
5) Clément, M.J-M., Alais, P., Roucayrol, J-C. and Perrin, J. (1982): Computerized ultrasonic tomography by electronic scanning and steering of a ring array. In: Acoustical Imaging, Vol. 10, Ed. P. Alais and A. Metherell, Plenum Press, New York, pp. 397-413.
6) Clément, M.J-M. and Alais, P. (1981): An electronically steered and scanned 450 elements ring array for transaxial ultrasonic tomography in breast imaging. Ultrasonics International 81, Brighton (U.K.), 2-5 july 1981 (Abstract).

REVIEW AND PERSPECTIVE OF BIOLOGICAL TISSUE CHARACTERIZATION

A. Bertini[1], A. Casini[2], G. Castellini[2], M. Cavallini[1], G. Ciocia[1], P. L. Emiliani[2], C. Lombardi[1], F. Lotti[2], L. Masotti[1] and S. Rocchi[1]
[1] Istituto di Elettronica, Facoltá di Ingegneria, Universitá degli Studi di Firenze, Italy
[2] Istituto di Ricerca sulle Onde Elettromagnetiche, Consiglio Nazionale delle Ricerche, Firenze, Italy

INTRODUCTION

The first studies in the field of tissue characterization were developed by our team in 1973. The work has since been oriented according to two main directions of research. The first concerned the problem of the optimization of spatial distribution of ultrasonic energy within the human body, in the examined field. Currently there are researches on the scanning of ultrasonic beams and on the programmable focalization. The second direction of investigation concerned the optimization of the ultrasonic pulse shape within the time-domain and the extraction and presentation of the echoes.

Transmission channel

The identification of a transmission channel for the ultrasonic signals has been our primary research. The channel is conceived as a series of time intervals to take into account the modification suffered by the transmitted signal S(t). The S(t) signal is transmitted by the probe, which has a complex transfer function T(f), through different layers of tissue, which have a transfer function Hi(f,x). The transmitted signal interacts with the interfaces characterized by the segment τ ,i,i+1, depending on the relevant transmission coefficients; and finally is received by the probe, with transfer function B(f).

Simulation

Utilizing the model outlined above, a simulation has been carried out using an IBM 1130 computer to evaluate the degradation suffered by an ideal signal with rectangular spectrum, for the way forward and back from the mitral valve. The obtained results, utilising the Chirp-Z transform, have shown a stretching in the behaviour of signal $S_r(t)$ with respect to S(t), yielding a 42% increment of spectral width at -3 dB, mainly due to selective absorption.

Time absorption coefficient measurement

The transfer function Hi(f,x) has been schematized, for the coefficient of absorption "α" in the form $\alpha(f) = Kf^m$ with K and m constants relating to the tissue. A measurement method has been conceived and realized to verify the tissue differentiation through the use of ultrasonic spectroscopy and to utilise the findings for design of echographic instruments.

Measurement method and system

The method is differential; during a sequence of measurements, the ultrasonic probe (corrected for the eventual squinting angle) is directed towards a spherical-cylindrical target. Subsequently the sample of biological tissue is interposed. This instrument has been designed and constructed to guarantee the reproducibility of the results. The mechanical system which couples probe and target is worked with a precision of 0.1 mm. The bath is filled with a 0.9% solution of NaCl in deionised, degassed water, at a temperature of $37 \pm 0.1^{\circ}C$; the sample of tissue is measured for 30 min. starting at 15 min. after the excision from the animal.

The measurement is carried out with a frequency synthetiser and a modulator coherent with the transmitted frequency, which produces a 30 µs tone burst. The pulse-spectrum was sufficiently narrow with respect to the appreciable variations of the transfer function of the system, to get significant samples, of the system transfer function, by changing the carrier frequency by steps of 100 KHz.

The data acquisition utilised a two channel fast sampler, BIOMATION mod. 8100, interfaced to a computer LABEN 70; the echo-signal was sampled on the first channel, on the other the carrier frequency to provide a consistent time-reference. The final measurements were obtained by averaging many subsequent samples pertaining to the same time and the final results are the averages obtained from many samples of tissue. Measurements were carried out on skeletal muscle, fat and blood. The m and K values were obtained by fitting a straight line to the parameters using the least squares method. Measurements of phase frequency dependence were performed on samples of blood. The described method has shown difficulties for direct "in vivo" application, although it could be interesting to examine the possibility of utilising ultrasonic spectroscopy for histological examination.

Realization of echographic instrumentation

During the period 1977-1981 the group, in cooperation with medical personnel and industry, has contributed to the formation of echographic instruments, according to a finalized research programme of the Italian Research Council. As a result of this cooperation there are now, on the market, a TM echocardiograph with programmable signal processing, a miniaturized linear array, a sector scanner with original electromechanical hardware. As prototypes we have devised the Eptacomp, a system for automatic compound scanning with the electromechanically moved head containing seven double ring transducers electronically switched according to an appropriate sequence, an instrument for the synchronous presentation of TM and real-time Doppler tracings, and a system for MTI ultrasonic imaging. Transducer characterization has been performed in cooperation with the Non Destructive Testing Laboratory of the CCR in Ispra. The grant obtained during the last few years permitted us to continue our researches on tissue typing with the aim of realising a dedicated instrument with optimised transmitted signals, linear array utilisation, digital control of the array and digital processing for reconstructing the ultrasonic field behaviour within the human body.

Ultrasonic field reconstruction with linear array

In order to characterize the tissues, one may resort to arrays or mosaics of transducers to avoid the phase cancellation effects over large apertures. The reconstruction of the pattern by means of an array can also give a better description of the macroscopic morphology, whose knowledge is necessary to improve the measurements. We think that the use of multi-element probes in experimental research is greatly eased by digital systems. The bandwidths and durations of the echographic signals are easily encompassed by modern analogue to digital converters and digital components. Thus it is possible to use digital methods to extract the information supplied by the transducers.

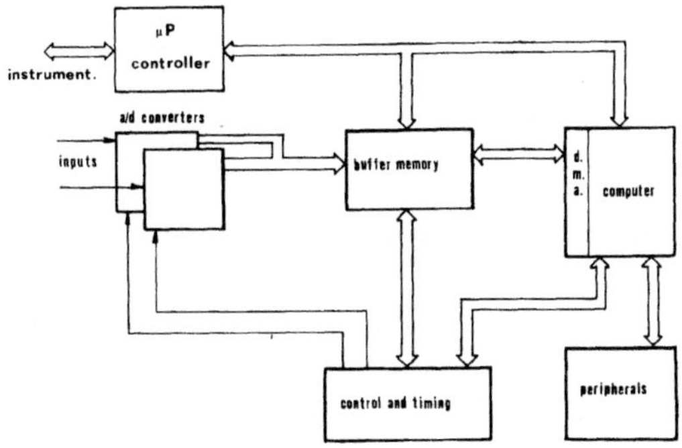

Figure 1. Digital Acquisition System

In our laboratory we work with a digital acquisition system, based on a general purpose minicomputer provided with an external memory of 5 Mbyte and several A/D conversion units. All the external equipment is controlled by microprocessor cards. One of the A/D converters has a conversion time of 50 ns and a dynamic range of 9 bits. Another ADC unit is equipped with 30 ns, 4 bit "flash" converters. The sequential access to the memory is achieved by interleaving the memory cards to reduce the write/read time. The system can act as buffer memory in the acquisition of ultrasound signals both for "in vitro" testing and "in vivo" measurements, with the aid of commercial equipment for monitoring. It can also serve as a refresh memory in the TV display of images. In its design we met with these problems :

- the high conversion frequency and the large amount of data,
- the exactness of the acquisition time (also in relation with target movements),
- the dynamic range of the signals,
- the digital noise (mainly due to the minicomputer).

In connection with the first point there is the problem of the amount of computation. This requires large memories and excessive computing time. Our sampling units are fast enough for ultrasound signals and, moreover, we are able to perform non-uniform sampling of the second order. One important thing to be mentioned is that algorithms necessary to coherently reconstruct the field can work independently on each series of non-uniformly sampled data and recombine the results at the end. So, the interpolating filter can work only on the lines to be displayed.

The exactness of the acquisition times is of capital importance in a digital acquisition system. Many factors must be controlled to reach the needed accuracy, such as the precision of the conversion frequency, the consistancy of the delays and the lack of the jitter by the transducer excitation circuitry. The timing of our system is stabilized by a free running crystal reference. The external ultrasound apparatus is also synchronised to this reference with a maximum jitter of two clock cycles (= 60 ns). To perform correct phase reconstruction the time of start of the ultrasonic pulse must be known within a margin which has to be small in comparison with one period of the carrier. (If the carrier is 3 MHz, 1/8 of a period is 48 ns). In our system the piezoelectric transducers are pulsed by capacitors switched by power MOSFET with excitation voltages greater than 100 V in order to get good S/N ratios. Usually we use MOS components with relatively low capacitance (50 pF) driven by CMOS, but when the voltages are higher (> 200 V) and the transducers' impedance are lower (annular arrays), we use an original, very compact, driver in which the MOS (700-1200 pF of input capacitance) is driven by a transistor in the avalanche mode. This device can be triggered with TTL pulses. The presence of the MOS devices avoides the occurence of spurious trigger pulses. It should be mentioned that the MOS has no avalanche sustaining voltage. The delay between the trigger pulse and shot is less than 25 ns, while the jitter (due to the thermal drift) is adequately small.

The movement of the regions where the ultrasound beam propagates is an important limiting factor, especially when arrays are used. "In vitro" it is always possible to electronically scan the elements of an array and later perform an "aposteriori" reconstruction of the field, but in an "in vivo" situation the collection time between the first and the last element must be small enough to avoid the target displacement being larger than a small fraction of the wavelength. To allow higher target speeds one should use more receiver elements in parallel and, consequently, more A/D converters. The peak speed of access to the memory would then increase. This is, in our opinion, the worst limit to the feasibility of a wholly digital echographic system. In order to have a low input noise we used an active multiplexer where each channel consisted of a low noise current amplifying FET and a feedback stage. This circuit was designed to maximize the S/N ratio per channel as a function of the number of channels. The full bandwidth of the receiving amplifiers is 20 MHz. Between 3 and 5 MHz one single channel has an equivalent imput of 5 μV; any successive channel introduces 1 to 2 μV more. The dynamic range is about 100 dB.

Computing time is a big problem in this numerical approach. To solve it we followed two approaches: one involves processing the acquired signals, the other the introduction of an "off line" dedicated processor, in parallel with the memory. For the first solution we designed a real time synthetic focus imaging system where the field of view is explored by pulsing one element of an array at a time and by recording the echoes received by the same element. This system allows an improved single line focussing, because the acoustic path between the element and the target is twice the corresponding physical distance, so that it may be seen as a single line focussing system with half wavelength. A mapping microprogram accumulated, on each point of any viewline, the samples which occur at a time which is equal to twice the travel time from any transducer element to that point. A quantitative analysis shows that the defocussing error (due to the limited bandwidth and the non-linear mapping rule) is acceptable. We also constructed a "scan converter" apparatus to convert video data from a generated compound B scan to standard TV raster. It was also designed to permit some on-line arithmetic operations on the single line of view. In the present version the arithmetic unit computes the memory address of the input sample, which is only stored in the memory if the content of the addressed cell is smaller than the sample itself. A cooperation is presently in effect with the Engineering Department of the Imperial College of Science and Technology of London to study the application of on-line digital filters by means of simplified arithmetic (power two). For the second solution ("off line" dedicated processor) we constructed a processor based on 4 bit slice chips (AM2901/3). In order to obtain fast computing times the microprogram does not permit branching, but a "condition" determines one of two parallel instructions. A fast auxiliary memory is included in the processor to increase the number of registers. A fast multiplier, the buffer memory and all the resources are mapped by an external dedicated microprocessor. The most severe limitation is the absence of a floating point arithmetic unit. The cycle is variable between 100 and 300 μs, depending on the instruction. The actual word length is 16 bits for the best use of a TRW multiplier.

Figure 2. Response of the synthetic focus system to a point target.

Discussion

A. BERTINI, A. CASINI, G. CASTELLINI, M. CAVALLINI, G. CIOCIA, P.L. EMILIANI, G. LOMBARDI, F. LOTTI, L. MASOTTI, S. ROCCHI

Review and perspectives of biological tissue characterization

Question asked by: J.M. Thijssen
What is your experience with annular array transducers and do you think they can be used in dynamic focusing for tissue characterization work?

Reply by: L. Masotti
I realized double concentric element transducers well working, mounted on my automatic compound scanning equipment (Eptacomp); a phase annular array consisting of seven annular elements and a central disk is now under development in our laboratory.
In my opinion a dynamic focused annular array should be the best transducer to be used for tissue characterization study; this, due to the fact that the beam has the same shape in each plane containing the acoustic axis. There are technological problems to solve in the construction phase to avoid spurious radial oscillation mode and acoustic and electrical coupling between the elements.

Question asked by: L.F. v.d. Wal
Added remark to question posed by J.M. Thijssen.
Main problem in manufacturing annular array's is the lack of knowledge of how the ring elements behave, i.e. which vibration modes are present in the piezo-electric material.

Reply by: L. Masotti
In manufacturing annular array's all parameters (dimensions, backing matching layers) should indeed be taken into account very carefully. We ourselves do not have an annular array as yet which is behaving satisfactory.

Question asked by: C.R. Hill

How do your values for attenuation coefficient in skeletal muscle (especially the exponent $n(\alpha=Af^n)$ compare with other published data (e.g. Nassiri et al., Ultrasonics 1979 or 80)?

Reply by: L. Masotti

We measured the amplitude attenuation coefficient of skeletal muscle excised from cows and calves maintained at $(37\pm0.1)^{o}C$ (temperature maintained by using 12 thermocouples distributed around the tissue sample at different depths).

The reflector was a spherical-cylindrical shape 15 mm in diameter.

The loss in energy due to the double interface tissue-physiological solution along the ultrasonic path from the transducer to the reflector and back was taken into account.

The bandwidth explored was from 1.5 to 3.0 MHz by using two squint angle corrected transducers, one centered on 1.8 MHz, the other one on 2.5 MHz.

Many measurements were carried out on each of the seven samples of tissue from different animals. The measurements were carried out with a frequency synthetizer and a coherent modulator producing 30 μs tone burst. The transfer function of the system was measured by changing the carrier frequency by steps of 100 KHz. The tissue sample was in the beam far zone.

Further discussions and remarks

The transmission measurements for the estimation of the attenuation were performed with a pulse echo method while using a spherical reflector. It was recommended to use a flat reflector instead, for the sake of avoiding diffraction effects in the estimation procedure.

A REVIEW OF INSTRUMENTS AND PROCEDURES DEVELOPED AT THE TECHNICAL UNIVERSITY OF DENMARK FOR THE STUDY OF ULTRASONIC TISSUE CHARACTERIZATION

Peter A. Lewin and Leif Bjørnø
The Industrial Acoustics Research Group, Technical University of Denmark,
Building 352, DK-2800 Lyngby, Denmark

Introduction

One of the primary objectives of medical ultrasonics is tissue characterization. Among tissue parameters that might be important for tissue characterization are reflectivity, acoustic impedance, propagation velocity and attenuation, including absorption and scattering. A fairly comprehensive review paper describing current status of tissue characterization including list of reports that have appeared in scientific journals up to 1980 has recently been published (1).

Instruments and procedures

Reliable measurement of propagation parameters such as attenuation and velocity requires appropriate ultrasonic probes that have good spatial and temporal resolution together with an adequate sensitivity to ensure acceptable signal to noise ratio during measurements. To this end we have developed miniature polymer (PVDF) ultrasonic hydrophones of superior performance. In comparison to traditionally used ceramic hydrophone probes they exhibit flat (to within $\pm 1,5$ dB) frequency response in the frequency range from 1 to 10 MHz. A comprehensive description of the ultrasonic hydrophone probes can be found in (2).
The construction of the probes allows their use, after suitable adaptation, for the assessment of tissue parameters in vivo (3).

A novel calibration procedure was developed to ensure absolute calibration of the ultrasonic probes in the frequency range between 1 and 10 MHz (4). The principal advantage of this method over conventional discrete frequency calibration is that it is able to detect rapid, continuous and significant variations in frequency response of the ultrasonic hydrophones due to e.g. spurious mechnical resonances. The method employs Time Delay Spectrometry (TDS) technique which not only facilitates assessment of hydrophone free field parameters in reflective environments but also allows both frequency response and directivity patterns of the ultrasonic hydrophone probes to de determined as a continuous function of frequency. Furthermore, this calibration procedure can be used to assess the effective area of the transmitter and receiver thus allowing diffraction corrections to be performed. These corrections are required for accurate determination of ultrasonic velocity and attenuation (5).

The carefully calibrated miniature hydrophones were used for measurements of velocity and attenuation as a continuous function of frequency for soft tissue in vitro (6). This work was performed to study the feasibility of measuring the acoustic field parameters in tissue with a miniature probe inserted via a biopsy needle. The measurement technique developed may be useful in tissue characterization work as it is expected that a variation in the attenuation of ultrasound through tissue as a function of frequency can be used as an useful index of the pathological state of the tissue.

Determination of another prospective index emerging from studies of finite amplitude effects in biological media, namely the second-order acoustic nonlinearity ratio B/A is the aim of current research, which is outlined in the next paper (7).

REFERENCES

1. Chivers, R.C. (1981): Tissue characterization. Ultrasound in Med. Biol. 7, 1-20.

2. Lewin, P.A. (1981): Miniature piezoelectric polymer ultrasonic hydrophone probes. Ultrasonics 19, 213-216.

3. Lewin, P.A. and Jensen, F. (1979): The use of miniature ultrasonic transducers for measurements in penetrable media. In Ultrasonics International '79, IPC Press, Guilford, UK, 279-286.

4. Lewin, P.A. (1981): Calibration and performance evaluation of miniature ultrasonic hydrophones using Time Delay Spectrometry. Proc. of 1981 IEEE Ultrasonics Symposium, Chigago, Oct. 1981 (in press).

5. Chivers, R.C., Bosselar, L. and Filmore, P.A. (1980): Effective area to be used in diffraction corrections. J. Acoust. Soc. Am. 68, 80-84.

6. Lewin, P.A. (1981): Focussed ultrasonic field distribution in tissue in vitro. In: Ultrasonics International '81, IPC Press, Guilford, UK, 434-439.

7. Bjørnø, L. and Lewin, P.A. (1982): On the use of the second-order acoustic nonlinearity parameter B/A for ultrasonic tissue characterization. (This volume).

ON THE USE OF THE SECOND-ORDER ACOUSTIC NONLINEARITY PARAMETER B/A FOR ULTRASONIC TISSUE CHARACTERIZATION

Leif Bjørnø and Peter A. Lewin
The Industrial Acoustics Research Group, Technical University of Denmark, Building 352, DK-2800 Lyngby, Denmark

Introduction

The nonlinear character of nature is for instance reflected in the fact that the fundamental mechanisms of sound propagation in liquid media are nonlinear (1). Nevertheless, a linearization of the governing equations has in a number of cases led to mathematically less complicated expressions with solutions showing surprisingly good agreement with experimental results. However, experimental evidence is now accumulating indicating that a linear treatment is not always sufficient for an exhaustive description of for instance ultrasonic wave propagation in biological media (2,3). In spite of the fact that earlier works showed no intensity dependence for ultrasonic attenuation in tissue, recent and more careful studies have proven the existence of intensity dependent attenuation as well as the formation of higher harmonics to finite-amplitude monochromatic waves in biological liquids and tissues (4,5, 6,7). The ultrasonic wave distortion course leading to the formation of the higher harmonics is due to two main sources, the material nonlinearity of the medium being expressed by the nonlinear character of its equation of state and the convection nonlinearity expressed by the fact that the local particle velocity is a function of the local pressure amplitude in the wave. Since the high frequency components of an ultrasonic wave are absorbed more readily than are the lower frequency components, the effective absorption of the distorted wave is greater than the absorption of a monochromatic wave of the fundamental frequency. This also leads to a higher rate of energy deposition within the propagating medium from a distorted wave. The widespread use of ultrasonic energy in medical diagnosis and therapy and the recent research showing the existence of nonlinear acoustic effects in biological media have led to a strongly increased interest in investigations aiming at providing enough information for an evaluation of the second-order acoustic nonlinearity parameter B/A as a tissue characterizing parameter on a par with attenuation, acoustic impedance, etc. (5,7).

Theoretical considerations

Several procedures can be used for an isolation of the second-order acoustic nonlinearity parameter B/A which characterizes the nonlinear relationship between pressure and density in a biological medium (1). These procedures have their advantages and disadvantages which will be discussed in this section. For lossless wave propagation Fubini solution (8) may be transformed to the following expression for determination of B/A from the second harmonic pressure amplitude P_2 as a function of source distance x

$$\frac{B}{A} = \frac{2\rho_0 c_0^3}{\pi f}\left(\frac{P_2(x)}{x P_1^2(0)}\right)_{xP_1(0)\to 0} - 2 \qquad (1)$$

where $P_1(0)$ is the peak amplitude of the fundamental at the source. ρ_0 and c_0 denote the density and the velocity of sound in the medium, while f denotes the fundamental frequency. Expression 1 was recently used for determination of B/A for several biological liquids (5,7), but the applicability of eq. 1 is limited due to the assumptions underlying the derivation of eq. 1. These assumptions include plane wave propagation, no influence of loss mechanisms and source distances small compared to the discontinuity distance, i.e.

the distance for the first formation of a discontinuity (shock) at the zero-crossing of the distorted wave.
The demand of plane wave propagation may to a certain extent be satisfied by measurements performed in the nearfield of an ultrasonic transducer only. However, nearfield measurements are complicated due to the existence of diffraction effects leading to a rather involved field pattern. Measurements performed using miniature hydrophones positioned on the acoustic axis of the transmitting transducer may allow for diffraction correction to be carried out in order to determine $P_2(x)$.
The demand of a source distance, which is consistent with the approximation forming the basis for eq. 1, namely, that only the second harmonic is considered and that series development of Bessel functions to second order terms only is performed, will be satisfied for nearfield measurements done in an ultrasonic field of not too high an initial amplitude $P_1(0)$ and considering the real medium attenuation effects leading to an increased discontinuity distance.
The most critical assumption in the derivation of eq. 1 is that the medium shall be lossless. The fact that loss mechanisms, i.e. absorption and scattering by inhomogeneities, are present in all real media render it necessary to take into account consideration of the attenuation effects in relation to ultrasonic wave propagation. A procedure developed by Thuras et al.(9) is based on the assumption that the attenuation of the fundamental and the second harmonic amplitudes are mutually independent and that the rate of change with propagation distance of the second harmonic amplitude is the sum of changes caused by the nonlinear generation of this second harmonic. These assumptions, which lead to good agreements with experimental findings for instance for plane finite amplitude wave propagation in gas-filled tubes (10), lead to the following expression for the second harmonic pressure amplitude as a function of source distance:

$$P_2(x) = \frac{(2+B/A)\pi f}{2\rho_0 C_0^3} \cdot P_1^2(0) \left[\frac{e^{-\alpha_2 x} - e^{-2\alpha_1 x}}{(2\alpha_1 - \alpha_2)} \right] \qquad (2)$$

from which B/A may be determined when the attenuation coefficients α_1 and α_2 for the fundamental and the second harmonic amplitude , respectively, have been determined experimentally. As shown by Dunn et al. (7) for $(\alpha_2 - 2\alpha_1)x \leq 0.5$, which may be obtained for small source distances x, a reduction of eq. 2 introducing an error less than 1% may lead to:

$$P_2(x) = \frac{(2+B/A)\pi f}{2\rho_0 C_0^3} \cdot x\, P_1^2(0) \cdot e^{-(\alpha_1 + \frac{\alpha_2}{2})x} \qquad (3)$$

which for known α_1 and α_2 leads to an expression for B/A given by:

$$\frac{B}{A} = \frac{2\rho_0 C_0^3}{\pi f} \cdot \left(\frac{P_2(x)}{x\, P_1^2(0)} \right)_{x P_1(0) \to 0} \cdot e^{(\alpha_1 + \frac{\alpha_2}{2})x} - 2 \qquad (4)$$

which leads back to eq. 1, now with introduced attenuation.
Another procedure, which via an experimental determination of the second harmonic pressure as a function of source distance x can form the basis for a calculation of B/A for biological tissue, is based on the perturbation solution originally developed by Keck and Beyer (11). The following expression for $P_2(x)$, which is made dimensionless by means of the source amplitude P_0

of the fundamental finite-amplitude sinusoidal wave with angular frequency ω may be expressed as:

$$\frac{P_2(x)}{P_o} = R\left[N(1-R) - \left(\frac{N^3}{4}\right)(1-R)^3\cdot(3+R) + \left(\frac{N^5}{16}\right)(1-R)^5\cdot(22+3R-5R^2-3R^3-R^4)\right] \quad \ldots\ldots (5)$$

where $R = \exp(-2\alpha x)$ and $N = \frac{\omega P_o}{4\alpha\rho_o c_o^3}\left(1 + \frac{B}{2A}\right)$

From eq. 5 may be seen that the calculation of B/A demands knowledge of the density ρ_0 and velocity of sound c_0 of the biological medium, and in particular a calculation of B/A demands a rather accurate determination of the fundamental wave, small signal attenuation coefficient α. This attenuation coefficient can be determined separately through careful measurements using miniature hydrophones in the ultrasonic transducer nearfield at a distance approximately equal to the last axial maximum. The measurements in relation to a determination of B/A from eq. 5 can be performed over the same nearfield distance in the biological medium, first using small signal amplitudes for the assessment of α and then finite amplitudes for the determination of $P_2(x)$. The use of eq. 5 does not make necessary an extrapolation back to x = 0 as was the case for eq. 1 and eq. 4, and only a minor path for the hydrophone measurements should be used. The original finite amplitudes emitted by the transducer need not to exceed 10^3 to 10^4 N/m^2, i.e. 180 - 200 dB/µPa, in order to establish the formation of the second harmonic. Due to the fact that α is the attenuation coefficient, i.e. it comprises both absorption and scattering effects, the use of eq. 5 will be advantageous for measurements in inhomogeneous media like for instance biological tissue. Recently, eq. 5 has been successfully used for calculations of finite-amplitude wave propagation in inhomogeneous materials like water-filled sediments (12).

A final procedure which does not demand the use of finite-amplitude waves for the experimental determination of physical parameters making a calculation of B/A possible is based on thermodynamical expression originally suggested by Beyer (13) based on an interpretation by Rudnick (14). The following thermodynamical expression can be used for a determination of B/A for an inhomogeneous medium (12) without consideration of attenuation influences:

$$\frac{B}{A} = 2\rho_o c_o\left[\left(\frac{\partial c}{\partial p}\right)_T\right]_{\rho=\rho_o} \cdot \left(\frac{2c_o T\beta}{C_P}\right)\left[\left(\frac{\partial c}{\partial T}\right)_P\right]_{\rho=\rho_o} \qquad (6)$$

In eq. 6 ρ_0 and c_0 denote the ambient density and velocity of sound of the medium, respectively. T and p denote absolute temperature and pressure, respectively, while β is the isobaric coefficient of volume expansion being expressed by:

$\beta \equiv (1/v)\cdot(\partial v/\partial T)_p$ where $v = 1/\rho$, c_p is the specific heat at constant pressure.

Determination of the derivatives of the velocity of sound c with respect to p and T for constant T and p, respectively, forms the most crucial acoustic measurement to be performed in the determination of the factors comprised by eq. 6. Traditionally, the major contribution arises from the first term in eq. 6.

Experiments to be performed

An experimental evaluation of the various procedures suggested above will have to be performed with the aim to assess which procedure will lead to the most reliable and reproducible B/A determination over a temperature range normally applicable to biological media.
Determination of B/A using eq. 1 and eq. 4 demands a very thorough transducer calibration involving already available procedures (15). Determination of the beam pattern of the transmitting transducer and detailed knowledge of its near-field structure will be necessary and could primarily be obtained in degassed water and glycerine before measurements in biological media are performed. The potential influence of diffraction effects on the formation of the second harmonic should also be assessed (16). The finite-amplitude source level should probably not exceed 180 to 200 dB re 1 µPa for continuous waves in order to avoid permanent cell damages due to e.g. microbubble pulsations or cavitation influences. For pulsed operation of the transmitter a somewhat higher source level can be used (17,18).
Several biological media should be studied. Starting with animal plasma and whole blood the study should comprise bovine or porcine liver and muscle fibres which should be studied both homogenized and whole over a characteristic temperature range. The degree of homogenization should be varied in order to study the influence of the tissue architecture upon B/A.
In particular the attenuation in the biological media should be carefully studied due to its influence on B/A determined using eq. 4 and eq. 5. The advantage of using eq. 5 instead of eq. 4 for the determination of B/A is that only one attenuation coefficient is necessary in eq. 5. However, the higher exponents of N and R in eq. 5, making necessary a numerical calculation, will probably, due to the more involved structure of eq. 5, reduce the accuracy of the B/A determination.
Both eq. 4 and eq. 5 require a rather accurate determination of density ρ_0 and sound velocity c_0 in the biological media, which can be obtained using standard laboratory procedures which can also be used for a determination of several of the thermodynamical parameters necessary for the calculation of B/A from eq. 6, including c_p and β.
Special experimental procedures have, however, to be developed in order to determine the derivatives of the velocity of sound with respect to pressure and temperature. Pressures in the range from $0.5 \cdot 10^5$ N/m^2 to $5 \cdot 10^5$ N/m^2 and temperatures in the range from 32°C to 42°C should lead to an assessment of the derivatives and thus the B/A values for various temperatures and for a constant pressure of 10^5 N/m^2.

Conclusions

When *in vitro* measurements of B/A for various biological media under changing conditions have been performed with an uncertainty satisfactory enough to indicate which experimental procedure should be used for the most reliable determination of B/A, a reply may be found to the question: "Will the second-order acoustic nonlinearity ratio B/A carry information which can be used for a characterization of biological media under various conditions"? When an affirmative answer is obtained, then an appropriate procedure for *in vivo* clinical determination of B/A may be developed.

References

1. Bjørnø, L.(1976): Nonlinear Acoustics. In: R.W.B. Stephens & H.G. Leventhal (Eds.) Acoustics and Vibration Progress, Vol. 2, Chapman & Hall Publishers, London.

2. Muir, T.G. and Carstensen, E.L. (1980): Prediction of nonlinear acoustic effects at biomedical frequencies and intensities. Ultrasound Med. Biol., 6, 345-357.

3. Carstensen, E.L., Law, W.K., McKay, N.D. and Muir, T.G. (1980): Demonstration of nonlinear acoustical effects at biomedical frequencies and intensities. Ultrasound Med. Biol., 6, 359-368.

4. Goss, S.A. and Fry, F.J. (1981): Nonlinear acoustic behaviour in focused ultrasonic fields: Observation of intensity dependent absorption in biological tissue. IEEE Trans. on Sonics & Ultrasonics, SU-28, (1), 21-26.

5. Law, W.K., Frizzell, L.A. and Dunn, F. (1981): Ultrasonic determination of the nonlinearity parameter B/A for biological media. J. Acoust. Soc. Am., 69, 1210-1212.

6. Carstensen, E.L., Becroft, S.S., Law, W.K. and Barbee, D.B. (1981): Finite amplitude effects as the threshold for lesion production in tissues by unfocused ultrasound. J. Acoust. Soc. Am., 70, 302-309.

7. Dunn, F., Law, W.K. and Frizzell, L.A. (1981): Nonlinear ultrasonic wave propagation in biological materials. Proc. of 1981 IEEE Ultrasonics Symposium, Chigago, Oct. 1981.

8. Fubini, G.E. (1935): Anomalie nella propagazione di onde acustiche di grande ampezza. Alta Frequenza, 4, 530-536.

9. Thuras, A.L., Jenkins, R.T. and O'Neil, H.T. (1935): Extraneous frequencies generated in air carrying intense sound waves. J. Acoust. Soc. Am., 6, 173-180.

10. Bjørnø, L. (1981): A study of the transition from linear to nonlinear wave propagation in gas-filled tubes. Proc. of Ultrasonics International 1981, 199-204, IPC Science and Technology Press Ltd., Guilford, UK.

11. Keck, W. and Beyer, R.T. (1960): Frequency spectrum of finite-amplitude ultrasonic waves in liquids. Phys. Fluids, 3, 346-350.

12. Bjørnø, L. (1977): Finite-amplitude wave propagation through water-saturated marine sediments. Acustica, 38, 4, 195-200.

13. Beyer, R.T. (1960): Parameter of nonlinearity in fluids. J. Acoust. Soc. Am., 32, 719-723.

14. Rudnick, I. (1958): On the attenuation of finite-amplitude waves in a liquid. J. Acoust. Soc. Am., 30, 564-569.

15. Lewin, P.A. (1981): Calibration and performance evaluation of miniature ultrasonic hydrophones using Time Delay Spectrometry. Proc. of 1981 IEEE Ultrasonics Symposium. Chigago, Oct. 1981.

16. Ingemto, F. and Williams, A.O. (1971): Calculation of second-harmonic generation in a piston beam. J. Acoust. Soc. Am., 49, 319-328.

17. Lewin, P.A. and Bjørnø, L. (1981): Acoustic pressure amplitude thresholds for rectified diffusion in gaseous bubbles in biological tissue. J. Acoust. Soc. Am., 69, 846-852.

18. Lewin, P.A. and Bjørnø, L. (1981): Acoustically induced shear stresses in the vicinity of microbubbles in tissue. J. Acoust. Soc. Am. (in press).

Discussion

L. BJØRNØ and P.A. LEWIN

Non-linear acoustics in tissue characterization

Question by: H.G. Trier
What are the connections between non-linear effects and cavitation in tissue, as both are leading to harmonics?

Reply by: P.A. Lewin
It should be emphasized that the non-linear phenomena discussed here are separate from cavitation, although the formation of shock waves may enhance other medium effects involving cavitation and acoustic streaming. Generally, cavitation in tissue is still an open question, however, there is some indication that the cavitation is unlikely to occur for the acoustic pulses of short duration, characteristic of ultrasonic diagnostic devices (P.A. Lewin & L. Bjørnø, Acoustic pressure amplitude thresholds for rectified diffusion in gaseous microbubbles in biological tissue. J. Acoust. Soc. Am. 69, 846-852, 1981).
Determination of the non-linearity parameter B/A can be performed according to several procedures and as mentioned during the lecture, not all of these procedures require knowledge or measurement of the second harmonic.
Also, our aim is to determine B/A in tissues not comprising bubbles. Bubbly liquids are highly non-linear, see eg.: L. Bjørnø, Acoustic non-linearity of bubble liquids, Proceedings of the IUTAM Symposium on Mechanics and Physics of Bubbles in Fluids, Pasadena, June 1981.

Question by: L. Masotti
How can you treat theoretically the possibility to extract significant information from acoustic non-linearity in non isotropic media like several biological tissues seem to appear?

Reply by: P.A. Lewin
One of the primary aims of our research is to develop a procedure which would allow B/A value to be determined by finite amplitude wave propagation over relatively short distances in the biological material. In this way it should be possible to consider the medium isotropic over these short distances.

Question by: C.R. Hill

Please say more about the physical nature of "B/A". How far can one expect B/A values to differ between different tissue types, and what variances might one expect in the measurements?

Reply by: P.A. Lewin

The second order acoustic non-linearity parameter B/A characterizes the second-order relationship between pressure and density in a medium. The basic analytical expression for B/A was derived by making use of Taylor series expansion and denoting appropriate coefficients in this expansion. Moreover, the B/A ratio can be related to various dimensionless thermodynamical parameters, the Rao's constant, Wada's constant, the Grüneisen parameter etc., which illuminate the molecular liquid structure producing information about the thermal and internal pressures, the intermolecular spacing, the anharmonic nature of the intermolecular potential etc., see K. Black & L. Bjørnø: A general thermodynamical procedure for calculation of acoustic non-linearity parameters of fluids, Proceedings of Scandinavian Acoustic Society Meeting, 1980, Abo, June 1980.

Turning now to the second part of your comment:

The non-linearity parameter B/A as a function of concentration for aqeous solutions of bovine-serum albumine considered as a tissue model (see, e.g.: W.K. Law, L.A. Frizzell and F. Dunn, Ultrasonic determination of the nonlinearity parameter B/A for biological media, J.Acoust.Soc.Am. 69, 1210-1212, 1981) was measured and found to lie in the range from approx. 5.3 to approx. 7.5.

Question by: D. Nicholas

Can you comment on the applicability of monitoring B/A values in tissues, where there is both high attenuation (compared to water) and velocity dispersion, when using a short acoustic pulse rather than continuous waves.

Reply by: P.A. Lewin

There is some evidence in the literature on the assessment of B/A value in biological media using short tone-bursts of 5 μs duration at the frequency of 3.44 MHz (e.g.: W.K. Law, L.A. Frizzell and F. Dunn, J. Acoust.Soc.Am. 69, 1210-1212, 1981).

Question by: J.M. Thijssen
Could you possibly give a verbal description of the B/A ratio and do the audience and myself a great favour?

Reply by: P.A. Lewin
Please cf. reply to Dr. Hill.

Further discussions and remarks

Further discussion on the non-linearity measure was devoted to an effort to explain the estimation techniques further. The B/A-ratio which is the ratio of the second order term to the first order term of the relation of pressure to density, can for instance be measured from the energy of the second harmonic if relatively long transmission pulses are used (Dunn et al.). These authors assumed a lossles medium.

COMPUTER- AIDED TISSUE CHARACTERIZATION FOR HUMAN EYE

D. Decker and K. M. Irion
Institut für Biomedizinische Technik, Universität Stuttgart, FRG

1. Introduction

A promising tissue differentiation supposes a redundant acquisition of information on the interaction of ultrasonic energy with the examined tissue. The appropriate signal is the RF-time-amplitude echogram. The analysis of these echograms requires computer-aided methods of evaluation because - apart from attenuation - only in this case tissue dependent changes in the time , frequency and phase domain of the echogram can be recognized.

From the beginning in the year 1973 and in cooperation with TRIER, LEPPER, REUTER from the Klinisches Institut für experimentelle Ophthalmologie der Universität Bonn the approach of our group was characterized by

- use of reference signals
- signal analysis by techniques of pattern recognition in the time as well as in the frequency domain
- averaging technique in the time domain
- tissue characterization and differentiation with sets of features, describing the structure of the tissue

For an objective tissue differentiation knowledge about the structure of the tissue to be characterized is of crucial importance. A distinction has to be made between the determination of the distance between the individual tissue boundaries, that means measurement of layer thickness and the description of internal structures.

This project was supported by a research grant from the Bundesminister für Forschung und Technologie, Bonn, FRG.

2. Characterization of thin tissue layers

To determine the distance of boundary surfaces we use the cross-correlation method. In this case the preprocessed signal is cross-correlated with a reference echogram obtained from a tissue - like reflector. The smallest layer thickness Δs_{min} which can be measured corresponds to the mean wave length of the ultrasonic pulse. For a centre frequency of 15 MHz, Δs_{min} = 60 µm is obtained, with an accuracy of 3%.

Table 1 contains the significant properties for the characterization of the examined tissue layers like distance and structure of the boundary surfaces as well as data of the spatial inhomogenity.

It should be pointed out, that the "reflectivity feature" can only be used with certain restrictions Within a single recording differences of - 3 dB (90% ...10% interval) may occur, within a class of tissue up to - 23 dB (90%...10% interval).

Tab. 1: Set of features for boundary surfaces

Boundary surfaces	Features	Intact. posterior wall	Macular* degeneration	Detached Retina	Choroidal* Haemangioma	Malignant Melanoma
I	occurence	100%	100%	100%	100%	100%
	spectral distribut.	SP	SP	SP	SP	SP
	centre freq./$_{MHz}$	12,0	12,0	10,5	12,5	13,0
	reflectivity/$_{dB}$	-20	-27	-28	-42	-32
	boundary-dist.I-II/$_{\mu m}$	133	100	128	0,90,120,160	122
II	occurence	100%	-	95%	-	100%
	spectral distribut.	SP	SP	SP	ST	ST
	centre freq./$_{MHz}$	14,5	14,0	14,0	16,5	15...18
	reflectivity/$_{dB}$	-20	-27	-28	-42	-32
	boundary-dist.II-III/$_{\mu m}$	100	78	60..95..120	210	90
III	occurence	50%	*	25%	100%	60%
	spectral distribut.	SP	SP	SP	FS	SP
	centre freq./$_{MHz}$	13,6	13,6	13,6	(15,0)	14,5
	reflectivity/$_{dB}$	-20	-27	-28	-42	-32
	boundary-dist.II-IV/$_{\mu m}$	540	450-870			260
IV	occurence	100%	100%			100%
	spectral distribut.	SP	SP			FS, ST
	centre freq./$_{MHz}$	14,0	14,0			17,0
	reflectivity/$_{dB}$	-20	-27			-32

* statistically not confirmed, N< 10 cases

Explanation spectral distribution: SP = specular surface
ST = scattering surface
FS = structured surface

centre frequency: centre frequency f_c of the spectral distribution with a nominal transducer-frequency of 15 MHz

reflectivity: in relation to a Kel-F-reference (15 mV ≘ 0dB)

3. Characterization of internal echos

If the tissue does not reveal a layer structure but consists of statistically distributed scatter elements (internal echos of tumours, bleedings etc.) it must be taken into account, that the spectral distribution of such an echogram is not only influenced by the frequency dependence of the tissue elements but also by the distance of the scatter elements. We therefore choose a special procedure to get a tissue-specific spectral attenuation, that will be explained in detail in the following paper. Table 2 shows the features for the description of the internal tissue of tumours and pseudo-tumours.

Tab. 2: Features of the internal structure of tissues. Reflectivity in relation to a Kel-F-reference (15 mV ≅ 0 dB)

Tissue / Scatter Elements	Choroidal Haemangioma	Exsudative Inflammation	Malignant Melanoma
Number per Window (2 μsec)	4 ... 7	2	5 ... 6
Distance (μm)	150 ... 280	450	260
Spectral distributon	FS	ST	ST, FS
Center frequency/$_{MHz}$	15.0	14 ... 17	17 ... 19
Ratio of amplitudes	~ 1	~ 0,5	~ 1
Reflectivity/$_{dB}$	-42	-16	-40

ST: Scattering surface
FS: structured surface

4. Conclusion

These examples demonstrate the working hypothesis of our research group which says:

A reliable characterization and differentiation of tissue demands an analysis in the time domain as well as in the frequency domain of the tissue echograms. Only the compilation of several features gathered from both domains of the signal may yield a meaningful tissue characterization. This concept of signal evaluation allows the objective determination of layers and the structure of tissues also with other organs.

5. References

/1/ Decker, D.; Epple, E.; Leiss, W.; Nagel, M. (1973) Digital Computer Analysis of Time-Amplitude Ultrasonograms from the Human Eye II. Data Processing, Journ. of Clin.Ultrasound, 1, 2 156-159

/2/ Decker, D.; Irion, K. (1979) A-mode RF-Signal analysis (frequency domain), Ultrasonic Tissue Characterization, Ed. J.M. Thijssen Stafleu's Scientific Publishing Company, Brussels.

/3/ Decker, D.; Trier, H.-G.; Epple, E.; Reuter, R.; Nagel, M.; Lepper, R.-D. (1980) Computer aided tissue differentiation in ophthalmology, Investigative Ultrasonology, Eds. C.R.Hill, C.Alvisi, Pitman Medical

/4/ Decker, D.; Irion, K.M. (1980) Examination of thin tissue layers, Fifth Int. Symp. on Ultrasonic Imag. and Tissue Charact., Gaithersburg, USA

/5/ Decker, D.; Faust, U.; Irion, K.M. (1981) Examination of Thin Tissue Layers in the Human Eye, Med.Progr.Technol. 8, 83-91

Discussion

D. DECKER, and K-M. IRION

Computer-aided tissue characterization for human eye

Question by: J.M. Thijssen
When you are talking about scattering elements, and talking about distance of elements and density of scattering centers, are you then talking about actual inhomogeneities in the tissue, or is it just a concept, a kind of term for convenience?

Reply by: D. Decker
We hope to extract from our values the real physical conditions. But I want to say clearly, that distance, density of the so-called scatter elements are not the realistic values.

Question by: D. Nicholas
Could you specify the reference spectrum used for your cross-correlation analysis.
Is your autocorrelation based upon the half maximum width of the first correlation peak, or does it represent a peak separation.
If the latter can you not also see minima within the spectral domain.

Reply by: D. Decker
The reference spectrum is extracted from the signal of a flat tissue-like reflector (contact lens material).
We use the peak separation, because this value is easy to handle in an automatic processing system. We then see a minimum in the spectral distribution but with thin layers and a given band-width we may have only one minimum and that is not enough to estimate the layer thickness correctly.

ESTIMATION OF THE ACOUSTIC ATTENUATION OF INTRAOCULAR TUMORS IN VIVO

K. M. Irion and D. Decker
Institut für Biomedizinische Technik, Universität Stuttgart, FRG

Digital methods of A-scan analysis using broadband RF-signals are presented. These methods are used for the characterization of tissue of the posterior wall of the human eye - in vivo. The following topics are discussed:

- data acquisition
- signal preprocessing, especially the detection of representative echograms
- signal processing methods and the extraction of features for the differentiation of the tissue.

From the time-amplitude echograms, statements can be made about the attenuation of ultrasonic energy in the tissue. For small tissue segments, especially in the eye, the analysis of high frequency broad-band RF-signals is necessary for the estimation of the attenuation coefficient and respectively the description of the internal structure of the tissue.

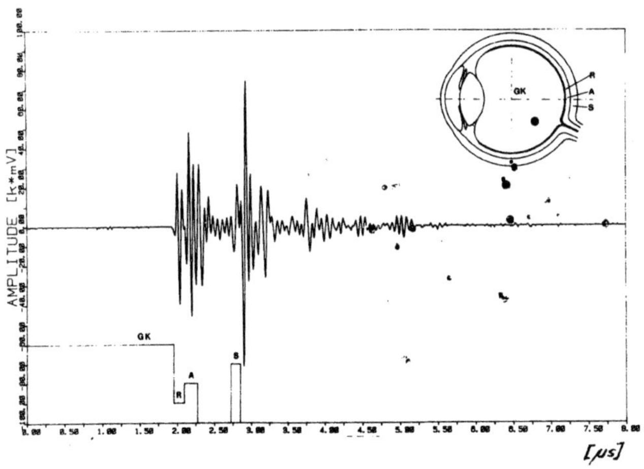

Figure 1.
Echogram of a normal posterior wall with retina (R), choroid (A) and sclera (S)

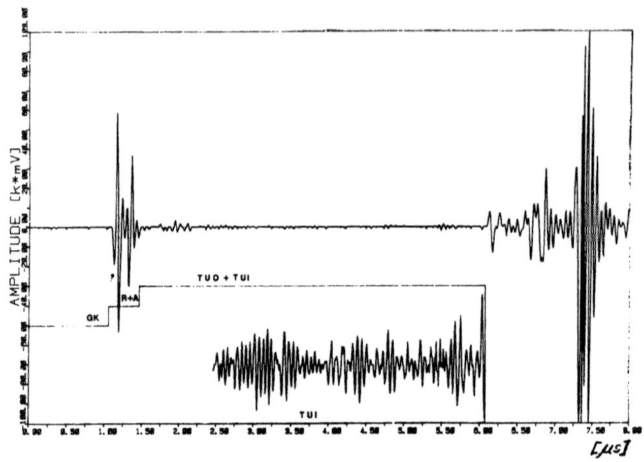

Figure 2.

Echogram of the posterior wall with a malignant melanoma in the choroid, - the retina and the sclera can be detected. The internal echoes of the tumor have small amplitudes. The tumor echogram is shown at larger scale below the original echogram. In this case the tumor has a length of 2.5 mm.

<u>Signal analysis</u>

I. <u>Data acquisition:</u>

- sample rate : 200 ... 256 MHz
- amplitude resolution : 8 Bit
- internal storage : 2 048 Byte
- repetition rate : ca. 4 Hz

II. <u>Signal preprocessing:</u>

- digital band-pass filter
- detection of quantization errors
- extraction of representative signals

III. <u>Signal analysis:</u>

a) <u>Extraction of features</u>

- attenuation coefficient (transfer function)
- determination of layer thickness (cepstral analysis, correlation)
- reflectivity (signal amplitude)

b) <u>Analysis of M-scans</u>

- phasecorrection
- detection of the layers (contour finding)
- estimation of the boundary distances as a function of time

Figure 3.

Overview of the signal processing and analysis being used for differentiation of normal and pathological tissues of the eye.

The sampling rate of the analog to digital conversion is maximal at 200 (256) MHz, at a resolution of 8 Bit. The transient digitizer has an internal storage of 2048 Bytes. The repetition rate is 4 Hz, depending on the transfer rate digitizer - storage medium of a minicomputer. The nominal frequency of the transducer is 15 MHz with an ultrasonic pulse length of 200 ns.
The preprocessing and the signal analysis is carried out on a minicomputer system (PDP-11/40), which is coupled via modem and emulator-software to the large scale computers CDC6600 and Cyber174 of the University of Stuttgart.

Preprocessing

At this high sampling rate the digitized echograms from the same location in the tissue are not identical. Disturbances may arise from the insufficient angle of incidence, from the pulsation of internal vessels or the movement of the operator. Therefore a class of about 30 scans will be digitized under the same conditions. Reducing computertime for the analysis of the whole quantity of echograms, a special correlation method detects representative signals from one class. For characterization, a few signal classes will be digitized from variable locations. Only the representative signals of classes and parameters of dispersion, which are available after signal selection, are used for the parameter extraction. The procedure of the signal selection is based on the crosscorrelation of n digitized signals of one class. A fast correlator calculates the n(n-1)/2 independent crosscorrelation functions of the class of signals. A peak detector estimates the maxima and the time delays of the crosscorrelation functions.
A special matrix operation, applied to the matrix of the maxima of correlation coefficients, detects the representative signal.

This detected signal is also the reference echogram for a phase correction of the other signals.

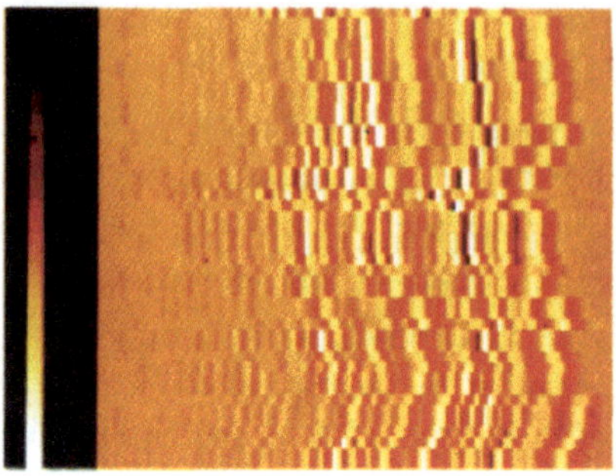

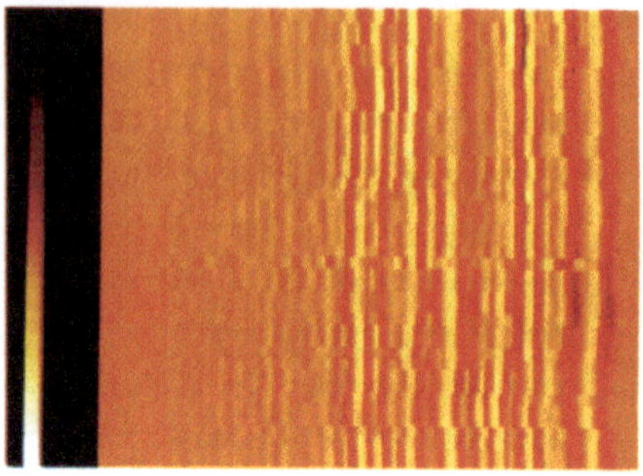

Figure 4:
Colour coded representation of the original digitized signals without phase correction - the horizontal line represents the depth of penetration.

Figure 5:
Signals with phase correction - the reference for phase correction is the detected representative signal. A phase coherent averaging can be done with signals of good correlation. This correlation method has the ability to detect and reject disturbed signals. Only the representative signal will be used for the parameter extraction.

Analysis of representative signals and parameter extraction

The differentiation of different types of liver tissue by the attenuation and the slope of attenuation gave practical results /4/. Several methods of the estimation were presented theoretically by Kak and Dines /3/ and by Kuc /4/.

In the works of Lizzi, Coleman /6/ parameters of attenuation were presented for tissue of the eye. The power spectrum, which was calculated from a 1.5 mm signal gate, was referenced to the power spectrum of the input signal. The input signal is obtained from a flat test-reflector. This estimate of a transfer function, however, does not represent the spectral behaviour of the signal gated from an internal echo. It only describes the spectral behaviour of the layers in front of the region of interest. Therefore the estimated parameters do not represent the attenuation as a function of depth of penetration. It is a parameter of reflectivity from a special location. Some familiar physical models which demonstrate typical relations of the attenuation of tissue layers are the following:

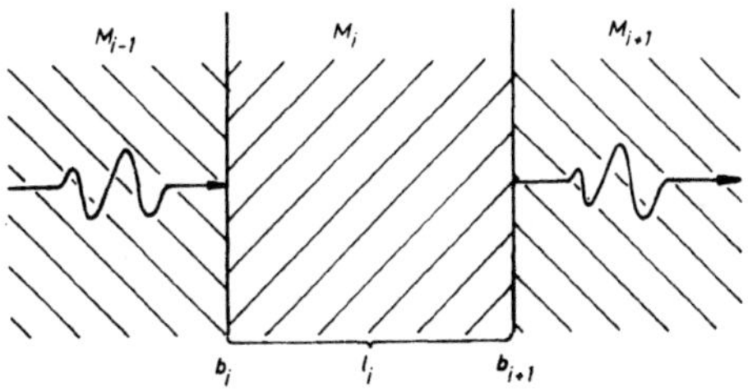

$$H_{i,T}(f) = b_i \cdot b_{i+1} \cdot e^{-\alpha_i(f) \cdot l_i} \cdot e^{-j\left(2\pi f \cdot \frac{l_i}{v_i}\right)}$$

$\alpha_i(f)$: attenuation coefficient of the ith tissue segment

l_i : thickness of the ith tissue segment

b_i : transmission coefficient at the boundary between the tissue segments M_{i-1}/M_i

Figure 6.

f : frequency

v_i: velocity in the ith tissue segment

Mathematical model for US-transmission for one homogeneous tissue layer without dispersion - this means that the group and phase velocities of the pulse wave packet are approximately equal. The transfer function is given by the presented equation and parameter definitions. αi (f) is the attenuation coefficient, depending on frequency and penetration depth li.

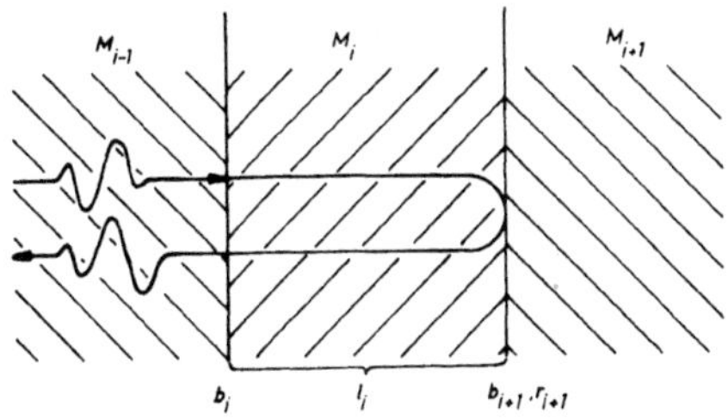

$$H_{i,R}(f) = b_i^2 \cdot r_{i+1} \cdot e^{-\alpha_i(f) \cdot 2 \cdot l_i} \cdot e^{-j\left(2\pi f \cdot \frac{2 l_i}{v_i}\right)}$$

with the same definitions (see transmission)

Figure 7.

r_{i+1} : reflection coefficient at the boundary between the tissue segments M_i / M_{i+1}

Model of US-reflection for one homogeneous layer without dispersion - the physical conditions are nearly the same as in the case of transmission. The transfer function for reflection is given by the presented equation. The difference between transmission and reflection is the double path length in the case of reflection. In both models the transmission and reflection coefficients are assumed to be constant factors.

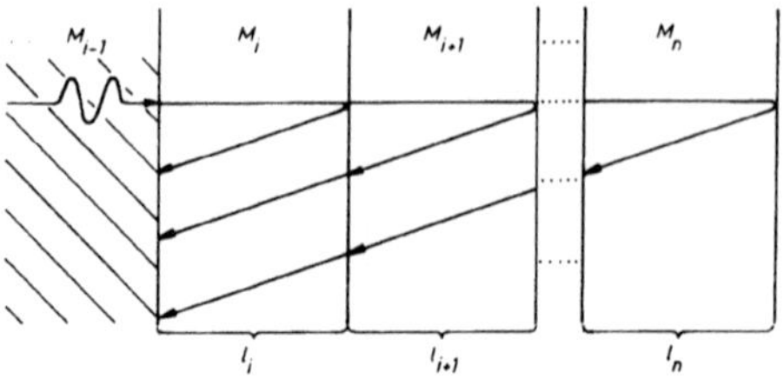

Transfer function for one tissue segment:

$$/H_{i,\,i+1}/ = B_i \cdot e^{-\alpha_i \cdot 2 \cdot l_i}$$

Transfer function for n tissue segments:

Figure 8.

$$/H_{1,\,n}/ = C \cdot e^{-\sum_{i=1}^{n} \alpha_i \cdot 2 \cdot l_i}$$

B_i, C : const.

Model of reflection for n tissue layers without multiple reflections. - The superposition of transmission - and reflection-model leads to a multilayered model of reflection. At the boundary surfaces of each layer the energy equation can only be assumed for the ideal case. On account of a rough layer surface and the non parallel layers the estimation of the transmission and reflection coefficients (bi, ri) in a multilayered tissue is not feasible when practically applied. The constant C can only be obtained as a measurable parameter, which presents a mean value of the reflection and transmission behaviour of the layers. Assuming a multilayered tissue with approximately equal attenuation coefficient αi, the attenuation coefficient $\alpha(f)$ can simply be calculated as the mean value of all αi.

$$/H_{i,\,i+1}(f)/ \quad = \quad B_i \cdot e^{-2\alpha_i \cdot l_i}$$

$$2 \cdot \alpha_i\,(f) \cdot l_i \quad = \quad \ln B_i + \ln / S_i\,(f)/ - \ln / S_{i+1}\,(f)/$$

S_i (f) : spectrum for ith received echo

S_{i+1} (f) : spectrum for i + 1th received echo

$$\alpha_i = \beta_i \cdot f \qquad f > 0$$

β_i : slope of attenuation (Np / MHz·cm, dB / MHz·cm)

Table 1.

$$2 \cdot l_i \cdot \beta_i = \frac{d\,(\ln / S_i\,(f) / - \ln / S_{i+1}\,(f) /)}{df}$$

The multilayered tissue is represented theoretically by a "sequence" of transfer functions Hi,i+1. Using this assumption the attenuation coefficients αi (f) can be calculated by the Fourier transform pair of the input- and output-signals Si(f) and Si+1(f). Many investigators have shown experimentally that in soft tissues the attenuation coefficient is linearly dependent on frequency in lower MHz-range.

With this linear relation the slope of attenuation β_i can be calculated. On account of the linear physical behaviour of the attenuation and frequency the differences of the logarithmic Fourier transforms can be approximated by a least square fit line. In the case of good correlation the slope of attenuation with the unit dB/MHz cm or Np/MHz cm is derived from the fit line. This is the theoretical background for the estimation of the attenuation. Problems will arise in practical use of the model, when the wave length (λ) is larger than the mean scatter dimensions (d) ($\lambda > d$). Interferences occur and in case of very small scatter dimensions (d) the scattering coefficient varies with the frequency with the power of four. For wavelengths in the range of scatter dimensions the scattering coefficient varies with the square of frequency. In this case the segmentation of the echoes out of the whole signal is not unique. A special algorithm for segmentation was presented by Roman Kuc for echograms of the liver. A modified algorithm for segmentation is used by our working group for the segmentation of short echograms of intraocular tumors and the posterior ocular wall.

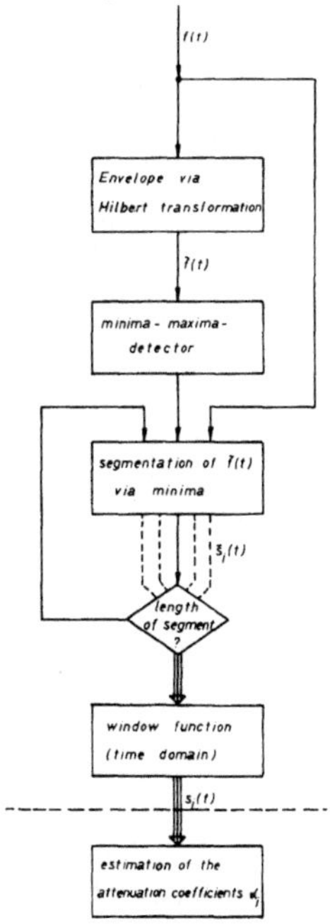

Figure 9.

Segmentation algorithm.

This segmentation algorithm differs from the algorithm presented by Roman Kuc in the calculation of the envelope and the segmentation using the minima.

An envelope contains peaks, which are caused by the reflection of a series of scatter elements. An algorithm detects minima and maxima of the envelope. The seperated nonoverlapping segments are limited by the location of the envelope's minima and the length of the reference pulse. Each segment is overlayed by a variable window function (e.g. Hamming, Hanning, Blackman, Bartlett). The window function, which is the best compromise of design criteria is the Hamming function (sidelobe supression).
Then the set of segments will be transformed by FFT and the amplitude and power spectra will be calculated. Attempting to reduce the number of echoes to one for spectral analysis leads to short data segments. Therefore the true spectrum of one echo is modified, or blurred, by shortening the length of the segments. The extent of adjacent frequencies contributing to any one point in the modified spectrum increases with decrease in the duration of time gate. Therefore we have to consider that there are not too short segments (≈ pulse length). In the 256-point data set the algorithm normally detects about seven segments in the case of malignant melanoma, four segments in the case of Macula degeneration, 3.5 segments in case of membranes. The attenuation and the slope of attenuation can be estimated from the sequence of the logarithmic amplitude or power spectra.

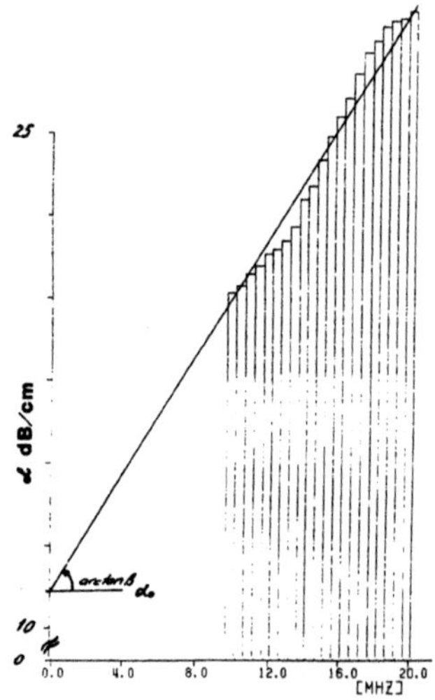

Figure 10.

Attenuation as a function of frequency of the normal posterior ocular wall. The fit line has a good correlation with the original curve. The parameters will be derived from this fit line. The slope of attenuation has values in the range of 0 - 2 dB/MHz cm.

Assuming the optimal distance of the scatter elements from the statistical model of Roman Kuc the variances of the values are still high. Only bleedings have a significant lower slope of attenuation β_1 compared to all cases of malignant melanoma. While the segmentation of the echograms is carried out under subjective criteria, different algorithms of segmentation are not comparable. In case of interferences, or Raleigh scattering the attenuation coefficient can only be estimated by averaging many coefficients at different depths of penetration.

A second formulation for the estimation of the frequency dependent attenuation is obtained by the short-time Fourier analysis. The short-time Fourier analysis is used to study variations of amplitude or energy of a signal as a function of time and frequency. The short-time spectra are usually measured according to two similar procedures (table 2).

Time Window Method:

$$S_T(t_1, f_1) = \left| \int_{-\infty}^{+\infty} s(t) \cdot W(t - t_1) \cdot \exp(-j2\pi f_1 \cdot t)\, dt \right|^2$$

$s(t)$: analytic signal

$W(t - t_1)$: window function centered at a time t_1

Frequency Window Method:

$$S_F(t_1, f_1) = \left| \int_{-\infty}^{+\infty} S(f) \cdot H(f - f_1) \cdot \exp(j \cdot 2\pi f t_1)\, df \right|^2$$

$S(f) \circ\!\!-\!\!\bullet\; s(t)$

$H(f - f_1)$: filter function centered at a frequency f_1

Table 2.
Short-time Fourier analysis time window method, - frequency window method

Time window method:
The analytic signal s(t) is multiplied by a window function w(t-tl) centered at a time tl. The short-time spectrum S_T(tl, fl), which is used in the time window method is defined by the first equation.

Frequency window method:
The signal is filtered using a transfer function H(f-fl) centered at a frequency fl. The short-time spectrum S_F is given by the second equation. It is necessary, that the filtering function should have an optimal resolution in time and frequency. But it is a common experience in the communication field, that one cannot simultaneously combine a function f (t) and its Fourier transform F(w) too accurately.

In practical purpose the window functions like Hamming, Hanning, truncated Gaussian can be considered as equivalent to an optimal filter function.

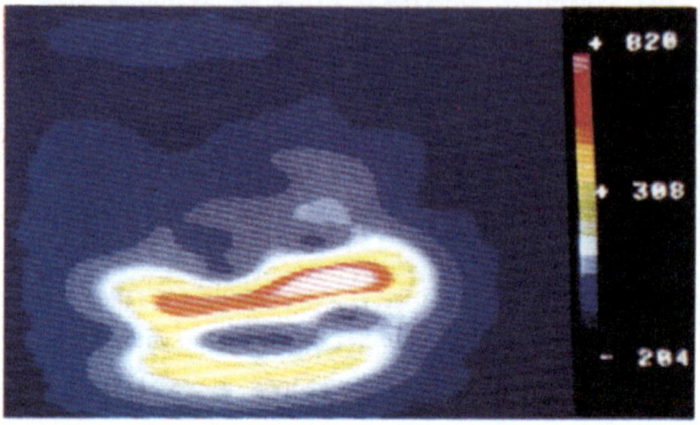

Figure 11.

Energy distribution as a function of time and frequency from the echogram of a detached retina - the vertical axis represents the time scale or penetration depth respectively with 2 µs. The horizontal axis represents the frequency scale with a maximum of 32 MHz and the colour represents the energy of the signal. Interferences occur at the point of maximal energy, because the windowed spectrum has no Gaussian distribution /2/. The estimation of an attenuation coefficient as a function of penetration length would fail for this short signal segment.

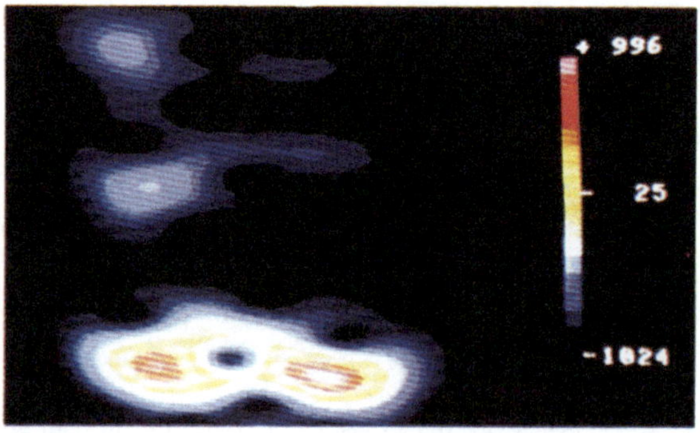

Figure 12.

Energy distribution of an internal echo of a tumor. Three maxima in the energy distribution can be seen. The first maximum indicates interferences. The shift of the center frequency is directly proportional to the slope of attenuation.

Difficulties will arise, if the spectra have no Gaussian distribution (interferences).
In this case the modified moving window method will achieve better results /5/. Instead of the shift of center frequency the shift of the center of gravity of the energy distribution is calculated.

However, the analysis of the two-dimensional energy distribution is not restricted to the estimation of the attenuation. A two-dimensional translation invariant transformation may give rise to a significant feature extraction.

References

/1/ Decker, D.; Epple, E.; Leiss, W.; Nagel, M.:
Digital Computer Analysis of Time-Amplitude Ultrasonograms from the Human Eye, II. Data Processing. -
Journal of Clin.Ultrasound 1, 156-159, 1973.

/2/ Decker, D.; Irion, K.M.:
A-mode RF-Signal analysis (frequency domain) - 1st. EEC Workshop on "In vivo Ultrasonic Tissue Characterization", Ultrasonic Tissue Characterization, Ed.: J.M. Thijssen, Stafleu Alphen/ Brussel, 231-244, 1980.

/3/ Kak, A.C.; Dines, A.C.:
Signal processing of broadband pulsed ultrasound: measurement of attenuation of soft biological tissues. - IEEE Transactions on Biomedical Engineering, Vol. 25 No 4, 321-344, Juli 1978.

/4/ Kuc, R.; Schwartz, M.:
Estimating the acoustic attenuation coefficient slope for liver from reflected ultrasound signals. - IEEE Transactions on SU, Vol. 26 No 5, 353-362, September 1979.

/5/ Kodera, K.; de Villedary, C.; Gendrin, R.:
A new method for the numerical analysis of non-stationary signals. -
Physics of the Earth and Planetary Interiors, Vol. 12, 142-150, Juni 1976.

/6/ Coleman, D.J.; Lizzi, F.L.; Jack, R.L.:
Ultrasonic spectrum analysis. -In: Ultrasonography of the eye and orbit. - Philadelphia: Lea & Fabiger, 83-88, 1977.

Discussion

K.-M. IRION and D. DECKER

Estimation of the acoustic attenuation of intraocular tumours in vivo

Question by: C.R. Hill
How much information would you expect to lose if you only recorded the envelope of the RF waveform?

Reply by: K-M. Irion
The envelope was calculated using the Hilbert transformation. The Hilbert transform has the same information as the RF signal. Deriving the envelope from the Hilbert transform of the RF signal, we will lose the phase information.

Further discussions and remarks

The characterization of multilayered membranes can be improved best by using the technique of crosscorrelation, while for the single layered membranes the cepstrum technique appears to be most suited.

The averaging of signals prior to applying signal analysis is performed with an auto/crosscorrelation and subsequent time shift of the signals.

ACTIVITIES ON ACOUSTIC IMPEDANCE TOMOGRAPHY AT L. M. A.

Jean-Pierre Lefèbvre
Laboratoire de Mécanique et d'Acoustique, Marseille, France

SUMMARY

Acoustic Impedance Tomography at the "Ultrasound" (US) Group of L.M.A., is the conjunction of two activities: the work on Acoustic Impediography of J.P. Lefebvre, and the work on Reconstructive Echo-Tomography of J.P. Sessarego. Acoustic Impediography is the quantitative extension of Echography first initiated by J.P. Jones. Reconstructive Echo-Tomography is an alternative to Echography, initiated by G. Wade. Because of its efficiency, we choose the technique of Reconstructive Echo-Tomography to solve the Three dimensional extension of Impediography.
We present here the general theory of the method, and partial experiments on one-dimensional Impediography and on the reconstruction of simple targets.

INTRODUCTION

Classical ultrasonic echography is a qualitative method for investigating biological media. It uses narrow-band ultrasonic pulses produced by a directional transducer, reflected by the various layers of the medium and collected by the same transducer which converts the acoustic pressure into an electric signal. Only amplitude information is used. It leads (when no multiple reflections occur) to the reflection power of the layers at a given travel time (the latter being interpreted as a function of depth). A new approach, adopted by Jones (1), is a quantitative one. It is based upon the use of broad-band pulses, which allows one to obtain the impulse response (by deconvolution) from the reflection by the medium. For weakly varying media, this impulse response is shown to be related to the acoustic impedance profile by a simple analytical expression. This simple expression is then inverted to recover the acoustic impedance profile from the measure of the impulse response. The generalization of the method for any media variables, for absorbing media, for three-dimensional structures

requires a complete reformulation of the problem in terms of an Inverse-Scattering Problem; this relates to Quantum Mechanics, where one may find advanced theoretical works on the subject (2).
The simplest approach, which is the one we used, is to transform the initial problem, which is an Acoustic one, into a Quantum Mechanical one, by the use of a Liouville-Green transformation. This approach was initiated by Ware and Aki (3) in Elasticity (Seismic prospection). The problem is then the recovery of the Scattering Potential from the reflection coefficient or from its inverse Fourier transform, the impulse response in reflection. This problem is solved for a wide class of potentials, by the general Faddeev-Marchenko method (2). Where it is impossible to use this general method, we propose approximate ones.

ONE DIMENSIONAL IMPEDIOGRAPHY : GENERALIZATION OF JONES' WORK (4)

Acoustic propagation in non absorbing heterogeneous media, when highly directional transducers are used, may be described by the one-dimensional Pekeris equation :

$$-\frac{\delta^2\Phi}{\delta t^2} + \rho c^2 \frac{\delta}{\delta x}\left(\frac{1}{\rho}\frac{\delta\Phi}{\delta x}\right) = 0$$

where t is the time, x the space coordinate, ρ the density, c the sound speed, and Φ the scalar acoustic potential.
This equation may be transformed into a Schroedinger one by use of a Liouville-Green transformation :

$$\Phi(x,t) \rightleftarrows \Psi(\zeta,t) \; : \; \begin{cases} \zeta = \begin{cases} \dfrac{x}{c_o} , x<0 \\ \displaystyle\int_o^x \frac{dx'}{c(x')} , x\geqslant 0 \end{cases} & (\zeta \text{ is the travel time}) \\ \Psi = \dfrac{\eta}{\eta_o}\,\Phi,\; \eta = Z^{-\frac{1}{2}} & (Z \text{ is the impedance } \rho c) \end{cases}$$

and of a Time Fourier Transform :

$$\Psi(\zeta,t) \rightleftarrows \psi(\zeta,\nu) \quad : \quad \psi(\zeta,\nu) = \int_{\mathbb{R}} \Psi(\zeta,t)e^{-i2\pi\nu t}dt$$

We have now to solve an Inverse Quantum Scattering problem, that is to compute the scattering potential $V(\zeta)$ for the Schroedinger equation :

$$4\pi^2\nu^2\,\psi + \frac{\delta^2\psi}{\delta\zeta^2} - V(\zeta)\psi = 0$$

where

$$V(\zeta) = \frac{1}{\eta}\frac{d^2\eta}{d\zeta^2}\,, \quad \eta = Z^{-\frac{1}{2}}$$

or

$$V(\zeta) = -\frac{1}{2}\frac{dq}{d\zeta} + \frac{1}{4}q^2 \quad , \; q = \frac{d}{d\zeta}\log Z$$

from the measure of the scattering coefficient $h(\nu)$ or its inverse Fourier transform $H(t)$: the Impulse Response in Reflection.
This problem (one dimensional, lossless) is well solved by the Faddeev-Marchenko method (2) :
The Scattering potential is given by : $V(\zeta) = 2\frac{d}{d\zeta}K(\zeta,\zeta)$, where $K(\zeta,t)$ is the solution of the Fredholm Integral Equation (of 2nd kind) :

$$K(\zeta,t) = -H(t+\zeta) - \int_{-\infty}^{\zeta} K(\zeta,\tau)\,H(t+\tau)d\tau \;;\; t \leqslant \zeta$$

The impedance profile $Z(\zeta)$ is then computed from $V(\zeta)$ by the substitution of : $\eta = Z^{-\frac{1}{2}}$ or $q = \frac{d}{d\zeta}\log Z$.
When using a successive approximation scheme, one may show that the first approximation is :

$$V^{(1)}(\zeta) = -2\frac{d}{d\zeta}H(2\zeta) \quad \text{that is } q^{(1)}(\zeta) = 4\,H(2\zeta)$$

which leads to :

$$\{\log\frac{Z(\zeta)}{Z_o}\}^{(1)} = 2\int_o^{2\zeta} H(t)dt$$

This is the Jones formula (1), which shows that our method is a simple generalization of Jones' Impediography.

EXTENSION TO ABSORBING MEDIA (5)

Biological tissues are dissipative media : they are the source of acoustic absorption-dispersion phenomena, especially in Impediography which needs very broad band pulses.

The simplest model that may be used is the viscous model, which leads to the acoustic propagation equation :

$$-\frac{1}{c^2}\frac{\delta^2\Phi}{\delta t^2} + \rho\,\frac{\delta}{\delta x}\left\{\frac{1}{\rho}\frac{\delta}{\delta x}\left(\Phi + \tau\frac{\delta\Phi}{\delta t}\right)\right\} = 0$$

where $\tau = \frac{\lambda+2\mu}{\rho c^2}$ is the dissipative parameter (time dimension) and λ, μ are the viscosity coefficients.

<u>In the general case</u>, one may show - after a Liouville-Green transformation - that this problem leads to an inverse quantum scattering problem with no solution at present :

The scattering potential $V(\zeta,\omega) = v(\zeta) + \frac{i\omega^3\tau(\zeta)}{1+i\omega\tau(\zeta)}$ is varying as the cube of the frequency.

<u>For weakly varying media</u>, one may use the classical scheme : approximate solution for the direct problem, followed by a simple formula inversion. By use of the travel time ζ (first part of a Liouville-Green transformation), the propagation equation becomes :

$$-\frac{\delta^2\Phi}{\delta t^2} + Z\,\frac{\delta}{\delta\zeta}\left\{\frac{1}{Z}\,\frac{\delta}{\delta\zeta}\left(\Phi+\tau\frac{\delta\Phi}{\delta t}\right)\right\} = 0, \quad Z = \rho c$$

We have now two parameters to identify : the acoustic impedance $Z(\zeta)$ and the dissipative parameter $\tau(\zeta)$.

Two mearurements are thus necessary to solve the inverse problem.

We choose the two impulse responses for reflection by the medium, the left one $H_L(t)$ and the right one $H_R(t)$, obtained by a double impediography approach.

For weakly varying media, one may calculate an approximate solution of the direct problem (Born approximation) and find the expression of $H_L^{(1)}(t)$ and $H_R^{(1)}(t)$. A simple linear combination of these expressions leads to :

$$q(\zeta) = 2\{H_L^{(1)}(2\zeta) - H_R^{(1)}(2\zeta_0-2\zeta)\} \quad , \quad q = \frac{d}{d\zeta}\log Z$$

and to :

$$\frac{d^2\tau}{d\zeta^2} - 2q\frac{d\tau}{d\zeta} + 2\frac{dq}{d\zeta}\tau = 8\{H_L^{(1)}(2\zeta) + H_R^{(1)}(2\zeta_o - 2\zeta)\}$$

where ζ_o is the "acoustic width" of the tissue sample.
These expressions then lead to the Impedance Profile $Z(\zeta)$ and to the Dissipative coefficient profile $\tau(\zeta)$, which is new information.

THREE DIMENSIONAL EXTENSION : IMPEDANCE TOMOGRAPHY (6)

The one dimensional model (except perhaps in Ophthalmology) is rather unrealistic : there are always biological structures with dimensions smaller than the beam width.
Neglecting absorption, one may start with the three dimensional Pekeris equation :

$$-\frac{\delta^2\Phi}{\delta t^2} + \rho c^2 \text{ div } \left(\frac{1}{\rho} \text{ grad } \Phi\right) = 0$$

with a supposed plane impulsive wave propagating in the direction $\vec{n}_o$:

$$\Phi_o = \delta(t - \vec{n}_o \cdot \frac{\vec{x}}{c_o})$$

In the general case, one may show - following Prosser's Work (7) - that the inverse problem needs the measure of the impulse response in Reflection in all space directions $\vec{n}_o$. This is quite unrealistic. However, we show that the computer aided reconstructive tomography method may be applied in a special case.

For weakly varying media, one may use itteratively the scheme for a Direct Problem Approximation Inversion.
With use of a vector travel time $\vec{\zeta}$ (acoustical path) defined as :

$$\vec{\zeta} = \begin{cases} \dfrac{\vec{x}}{c_o} & \text{in the homogeneous space } \mathbb{R}^3 - \mathbb{R}^{+3} \\ \displaystyle\int_{\vec{o}x} \frac{d\vec{x}'}{c(x')} & \text{in the homogeneous space } \mathbb{R}^{+3} \end{cases}$$

one reaches the equation :

$$-\frac{\delta^2\Phi}{\delta t^2} + Z(\vec{\zeta}) \text{ div}_\zeta \left(\frac{1}{Z(\vec{\zeta})} \vec{\text{grad}}_\zeta \Phi\right) = 0$$

with the excitation :

$$\Phi_o = \delta(t-\vec{\eta}_o.\vec{\zeta})$$

One may then write the equivalent integral equation, and solve it by an iterative method (Born). The Impulse response for reflection in the $\vec{\eta}_o$ direction $H_B(\vec{\eta}_o,t)$ is shown to be of first order :

$$H_B^{(1)}(\vec{\eta}_o,t) = \frac{1}{4}\ \frac{\delta^2}{\delta\zeta_1^2}\ \{\int_{\mathbb{R}^2} \log\frac{Z}{Z_o}\ d\zeta_2 d\zeta_3\}_{\zeta_1 = t/2}$$

where ζ_1 is the coordinate along $\vec{\eta}_o$, and ζ_2, ζ_3 the orthogonal ones.
So, the second integral of the impuls response for reflection in the direction $\vec{\eta}_o$ is the perpendicular projection of the Logarithmic Impedance Profile Log (Z/Z_o) in this direction. That is the exact formulation of the proposition stated by Wade (8).
So, a reconstructive tomography method, like the Sessarego one (initiated for simple targets in water) may be used (9).
If $P_\theta(r)$ is the projection in the Θ direction of the quantity $f(x,y)$ to be reconstructed; then

$$f(x,y) = \int_o^\pi \{\frac{\delta P_\theta(r)}{\delta r} * \frac{1}{r}\}\ d\Theta$$

is the reconstructive algorithm.
A simple approximation is shown to be the back-projection summation method.
It leads to :

$$g(x,y) = \int_o^\pi P_\Theta(r)\ d\Theta \quad , \quad g(x,y) = f(x,y) * \frac{1}{\sqrt{x^2+y^2}}$$

which is a good approximate for qualitative imaging, but surely insufficient for quantitative imaging. However, it is a good method for the study of the space time sampling effects.
All these algorithms are efficient for two dimensional problems, although they need to be adapted for the three dimensional situation.

EXPERIMENTS

The experimental program is conceived for both Medical Ultrasonic applications and Seismic Prospection applications. In the latter case laboratory experiments are considered as reduced models of the actual problem.

Partial experiments were made at scale 1/1 for 1-D Impediography and at scale 10/1 for Reconstructive Echo-Tomography. This last scale requires large mechanical equipment (1.5 m rotating arm), which allows an easy control of target positioning and the use of low frequency ($f < 1$ MHz) and low cost devices (analog to digital converters). It is a good academic model, and a step towards full scale 1/1 experiments.

We only recently bought a high performance A/D converter (100 MHz sampling rate) : the Biomation 8100. Before this, we only had a low frequency converter (1 MHz) sampling rate), which was efficient for Reconstructive Tomography experiments at scale 10/1, but not for Impediography. The experiments on 1-D Impediography were achieved at the "Centre d'Etudes des Techniques Chirurgicales (C.E.T.C.) de l'Hôpital Broussais" (Dr. Peronneau) with the Biomation 8100 of Herment.

Experiments on 1-D Impediography

We tested the simplified Jones' procedure on simple models (plates in water) at the C.E.T.C., using broad-band transducers conceived by the L.M.A./US (Garnier, Holtzer). These transducers are patented models (from ANVAR-CNRS) now constructed by Sodelec.
Figure 1 shows a result with a water-plexiglass interface and with the transducer T.30.1 (40 KHz to 3.7 MHz, total bandwidth at 25 dB), driven by a short excitation pulse.

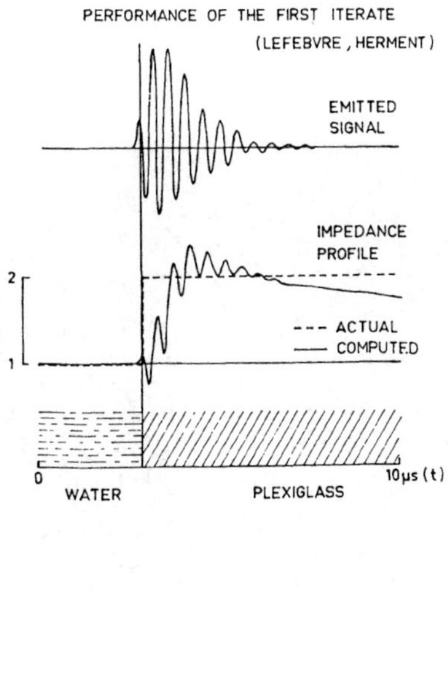

Figure 1. Experiments on 1-D Impediography at CETC (with Herment's chain and T.30.1).

The water plexiglass interface (Heaviside's Profile).

These experiments show the necessity for a frequency band extended towards the low frequencies (most of the general information comes from this zone).

We are now conducting the experiment at L.M.A., with our Biomation 8100. We intend to work with stratified media constructed by Société Nationale Elf-Aquitaine (S.N.E.A.) for seismic simulations. We conceived for this a high performance emission-reception chain, centered on the T.30.1, which allows us to detect reflectivities of about 2.5×10^{-5}, i.e., impedance variations of the order of 5×10^{-5}.

Experiments on Reconstructive Echo-Tomography

The experiments were made at low frequency (1 MHz, 10 μs detected wave-trains) with large axi-symetric targets (to provide a two dimensional problem) immersed in a large tank.

A rotating arm of 1.5 m allows us to measure the impulse response for reflection by the targets (completely insonified by the transducer) at all angles (multiples of 2°). The analog to digital conversion is now performed by the Biomation 8100. All operations are driven by a MINC 11/03 with floppy disk storage, pulse emission, retrodiffusion sampling. The Image Reconstruction and processing is made by a PDP 11/60 connected to the MINC.

Figure 2 shows a result, (from Sessarego), for an hexagonal steel bar, using only 32 projections (simplified back-projection summation method). The experiments show the important role played by the third dimension which is not taken into account (i.e., the necessity to have a rotation axis parallel to the target axis).

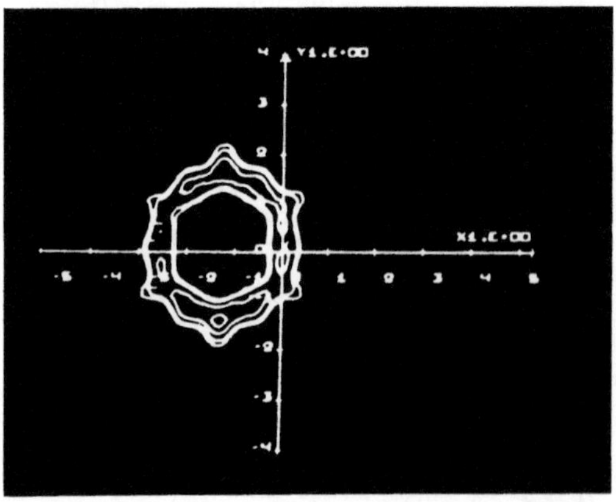

Figure 2. Experiments on target imaging by Reconstructive Echo-Tomography at L.M.A. (32 projections).

2,4 cm hexagonal steel bar.

FURTHER WORKS

We plan to construct smaller Reconstructive Echo-Tomography experiments, so that all experiments on Acoustic Impedance Tomography will be made at scale one, (between 40 kHz and 3.7 MHz, with T.30.1). These experiments will be considered at scale 1/10,000 in time and 1/24,000 in space, for seismic simulations using plastic materials (velocity scale : 1/2.4). The studies will then concern the various numerical procedures involved: deconvolution, inversion, image reconstruction and image processing.

REFERENCES

1. Jones, J.P. (1977): Ultrasonic impediography and its applications to tissue characterization. In: Recent Advances in Ultrasound in Biomedicine, Vol. 1, pp. 131-156, (D.N. White, Ed.), Research Studies Press, London.
2. Chadan, K. and Sabatier, P.C. (1977): Inverse Problems in Quantum Scattering Theory, Springer Verlag, New York.
3. Ware, J.A. and Aki, K. (1969): Continuous and discrete inverse scattering problems in a stratified eleastic medium. I. Plane waves at normal incidence. J. Acoustic. Soc. Am., vol. 45, nr. 4, pp. 911-921.
4. Lefebvre, J.P. (1978): Théorie d'une méthode quantitative d'investigation des milieux stratifiés : l'impédiographie acoustique. Acustica, vol. 41, nr. 1, pp. 11-20.
5. Lefebvre, J.P. (1982): Theoretical basis of an ultrasonic investigation method of stratified absorbing media. Acoustical Imaging, vol. 10, pp. 287-293 (P. Alais and A.F. Metherell, Eds.).
6. Lefebvre, J.P. (1981): Quelques méthodes quantitatives d'investigation par ultrasons, problèmes de modélisation et d'inversion de la rétrodiffusion. Dissertation, University of Provence/Aix-Marseille I.
7. Prosser, R.T. (1969 and 1976): Formal solutions of inverse scattering problems (I and II). J. Math. Physics, vol. 10, nr. 10, pp. 1819-1822, and vol. 17, nr. 10, pp. 1775-1779.
8. Wade, G. (1980): Acoustic echo-computer tomography. Acoustical Imaging, vol. 8, pp. 565-576 (A.F. Metherell, Ed.).
9. Sessarego, J.P. (1981): Reconstruction d'une cible à partir de mesures acoustiques, 8ème Colloque sur le Traitement du signal et ses applications (GRETSI), Nice, vol. 2, pp. 697-702.

Discussion

J.P. LEFEBVRE

Works on acoustic impedance tomography at L.M.A. (Laboratoire de Mécanique et d'Acoustique).

Question by: J.M. Thijssen
Is what you called in the beginning the first iteration step identical to what you later on termed a first order approximation?
Do you have a good criterion to stop the iteration?

Reply by: J.P. Lefebvre
Yes. One may show with a perturbution method that each iteration step of an integral equation is an approximation of the same order. So the two terminologies are equivalent.
About a criterion to stop the iteration, it is impossible to find a good one without experiments. All iteration procedures are noise-sensitive; they "explode" for high orders and no theory is able to say when. However, we think that an order of 5 iterations will in general be possible and sufficient.

Question by: D. Nicholas
Your detailed analysis utilises the 3-D wave-equation and absorption together with multiple reflections. However your eventual experiments are done in 1-D and Born approximation. How necessary is it in these circumstances to invoke such a detailed analysis. Do you in reality gain anything by these mathematical improvements?

Reply by: J.P. Lefebvre
The experiments are only at the beginning. We are now working to set up higher order iterates on experiments. We show the gain by using our mathematical improvement with numerical simulations, and the comparison is good with other more sophisticated - and more expensive - methods (optimal control theory). Perhaps the question would be: does anyone need for quantitative results? If yes, I think that there is no other solution than more Physics and more Mathematics (and Informatics).

Further discussions and remarks

The plans indicated to estimate three dimensional impediograms were further elucidated. The 3 D pictures are constructed from series of 2 D impediograms.

TISSUE CHARACTERIZATION IN OPHTHALMOLOGY

H. G. Trier and R. D. Lepper
Klinisches Institut für experimentelle Ophthalmologie der Universität Bonn, FRG

Introduction

The background of the ultrasonic laboratory in a clinical research entity, and the main activities in the past and presence are shortly reviewed. The actual activities include : Pulse-echo- and Doppler-methods; research and development (hardware and software) in tissue characterization, biometry, calibration/standardisation of equipment; clinical application and evaluation of the results; and ultrasonic therapeutic and surgical equipment.
The following summarizes our work in tissue characterization in the eye and orbit.

1. Clinical needs for tissue characterization in the eye and orbit.

In ophthalmology, for many years clinical echography has included routine techniques of tissue characterization, especially in A-mode work. For this purpose calibrated equipment with specific system characteristics, together with quantitative visual evaluation of the reflectivity of boundaries and of the echo pattern was early introduced.
Specialised tissue characterization techniques, partially overlapping with "biometry", are generally of interest, where conventional A- and B-mode echography fails, and after alternative diagnostic methods have been considered.
In the eye and orbit there is a need for such techniques, especially in the following conditions :

- in identifying the nature of tumours by direct approach, or by indirect characterization of the nature of surrounding tissues (tumour infiltration);
- in the assessment of tumour growth and size;
- in detection of vitreoretinal diseases, mainly retinal detachment by direct differentiation of the membrane in question, or by indirect approach at the wall of the eye;

- for improved detection and evaluation of small foreign bodies;
- for improved ultrasonic biometry of the optic nerve, the ocular muscles, and the "micro-anatomy" of the eye wall structures;
- for evaluation of the conditions of the retino-choroidal and scleral layers in various diseases.

2. Methodical approach.

The group has been engaged for ten years in the approach of tissue differentiation by computer-aided empirical analysis of the A-mode and RF-signal. Since 1973, we specialised on the RF-signal, because of its higher information potential. After experiments in vitro an optimized system was adapted to the living eye and has been under clinical evaluation since 1977 at the University Eye Clinic in Bonn to collect statistical data*).

2.1 Signal acquisition (Bonn).

2.1.1 Technical data of the system.

The region of interest in the eye is examined by a hand-held transducer, Ø 1/2", with water stand-off, focused at 50 mm, with f_c = 15 MHz, relative bandwidth > 0.3, pulse length 200 ns (- 10 dB). Receiver bandwidth 5-30 MHz (- 3 dB), input noise level < 15 µV eff; dynamics 76 dB. Time window a) 2.0, b) 10 µs. AD-conversion: a) 1974 till 1980 Tektronix R 7912 (double ended signal storage tube) with 9 bit (Y), sampling rate 256 MHz (equiv.), b) since 1981 Tektronix AD 3020 (7612), 8 bit (y), sampling rate 200 MHz. Data intermediate storage and transfer on tape/disc via minicomputer. The system characteristics are evaluated by reference signals.

2.1.2 Comment: Instrumental prerequisites for ultrasonic tissue differentiation.

For tissue characterization on the basis of RF-echograms several conditions have to be fulfilled. In a way of logical succession the different components of an ultrasonic system shall be described.

---

* With the support of BMFT/DFVLR; Grants no. 01 VI 057-ZA/NT/MT 224 a (Bonn) and 01 VI 047/ZA/NT/MT 224 a (Stuttgart).

Beginning with the generator, it should be assertained that the latter produces a pulse shape with one very steep and one very flat slope. Several commercial instruments show unsuitable oscillating pulse shapes, in which case the frequency of these electrical oscillations does not depend on the transducer.
The next link in the chain is the transducer. Some general criteria for evaluating the usefulness of transducers for tissue characterization are available; however, a minimum of requirements cannot be given very precisely. Required is a focused transducer, whose length of the focus zone should correspond with the depth of the lesion to be examined. A criterion with regard to the rotation symmetry is considerably more difficult to define. The iso-amplitude lines should be similar to concentric circles. For our ophthalmic technique the 6 dB-range does not exceed the maximal diameter of 0.7 mm. This was postulated because of the fact that the sectional areas of the tissue lesions to be examined are usually very small, and the surrounding tissue should not influence the RF-echograms. In our opinion, a transducer beam plotting system for quality assurance of the transducers is indispensable.
A point to point measurement can be performed either by hydrophones, or by means of a spherical test reflector with suitable diameter. We prefer the measurement with reflectors, because this approach is more similar to the conditions in the tissue. Moreover, appropriate hydrophones that are calibrated up to 25 MHz are scarcely obtainable.
Fig. 1 demonstrates as an example a sound field, measured by means of a reflector. This reflector was particularly shaped in order to avoid interfering backwall echoes.
Fig. 2 indicates a typical transducer defect: radial cracks in the plastique lens of the transducer are clearly distinguishable. It is not yet clarified, whether the aging of the plastique lens is in this case quickened by ultrasound. The cracks lead to water penetration in the crystal, and to changes in its performance.
The echo amplifier has to be considered as another important link of the system. Signal influencing or signal processing components are already incorporated in the usual ultrasonic equipment, as these systems mostly have to provide visually interpretable signals on the screen. For tissue differentiation a pre-amplifier with extremely high dynamics, and very low input noise should be used. In the first stage we used an electron tube. To get an optimal signal-to-noise ratio the attenuator was built behind the first stage.

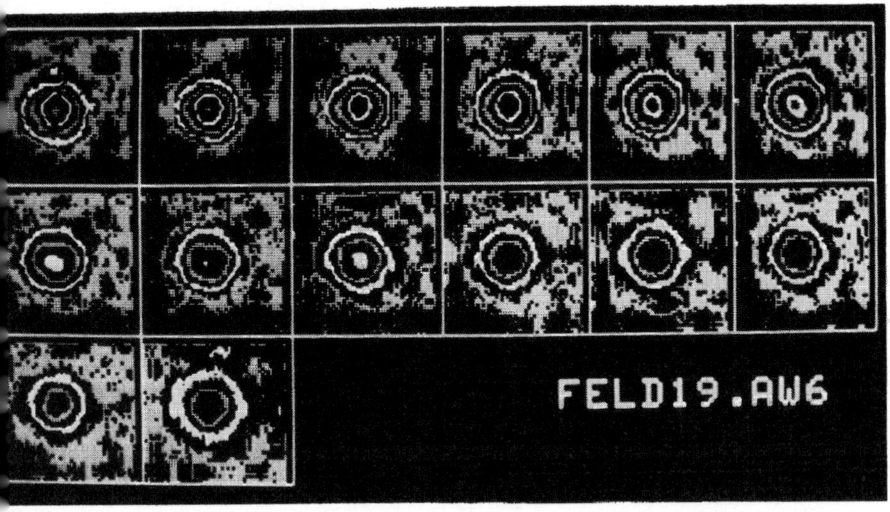

Figure 1. 14 maps of a sound field perpendicular to the transducer axis, being 1.5 mm apart.
The area of each plot is 3.9 x 3.9 mm^2. The iso-amplitudes drawn correspond to the absolute maximum, 1/2, 1/4, 1/8 and 1/16 of the maximum amplitude.

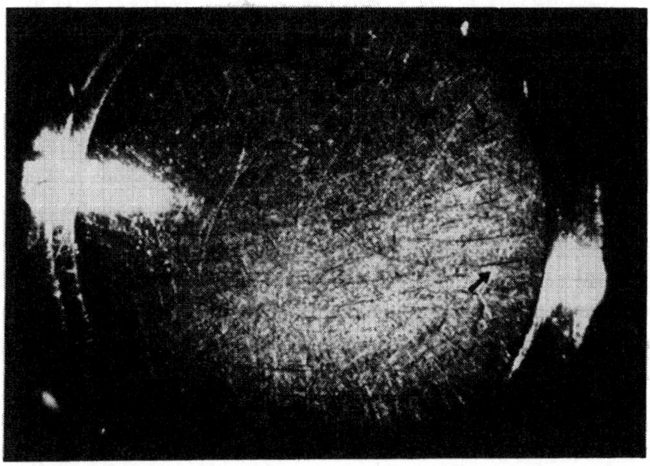

Figure 2. The photograph of the front end of a transducer demonstrates radial cracks in the plastique lens of a LSH transducer. Due to the corresponding water contact the hygroscopic piezo plate was destroyed.

A decisive problem in computer-aided tissue differentiation is the digitization of RF-echo-signals from the inner eye. As frequencies up to about 25 MHz may partly be present in the echo signals, the sampling rate should be higher than 50 MHz, theoretically.
In the initial period of our experiments not sufficient information concerning the frequencies involved, or their harmonics, respectively, was existent. A virtually unlimited system with respect to bandwidth was, therefore, necessary. The Transient Digitizer R 7912 (Tektronix) was the first analog/digital converter we used.
This system does not work in the manner of a classical A/D converter, but represents a direct combination of oscilloscope-tube and TV-camera. Therefore, simple noise reduction techniques by means of a graphic signal storage on the target of the tube system result.
On the other hand, the limited number of points provided by the R 7912 device is disadvantageous; it allows the digitization of very small timegates only, and needs a quiet and cooperative patient.
To evade these difficulties we tried to find alternatives. The device Biomation 8100 proved unsuitable for our purposes, because considerable digitization errors occurred. Recently, we replaced the R 7912 by the 7612 AD-system (Tektronix AD 3020), which works on two channels. Each channel comprises 2048 samples with 8 bit each, the sampling rate with 200 MHz maximum satisfied our needs. Using the 10 μs gate, identification of macroscopic structures is greatly improved as well as the influence of patients movements on the quality of the echograms, which is markedly reduced.

2.2 Signal processing (Stuttgart; see contribution K.M. Irion and D.Decker).

By external computer (PDP 11/40), using semi-automated procedures, software with modular structure, allowing interactive corrections under display control. Evaluated bandwidth 5-22 MHz. Preprocessing includes signal quality assurance and formation of one or more file-specific representative echograms. For this purpose from m signals of one examined region the $\binom{m}{2}$ combination of cross correlation functions are calculated, the representative signals are then determined by a matrix method. Signals with a high correlation with the representative signals are averaged in the time domain and used for a phase correction.
Processing is done in the time and frequency domains empirically, leading to sets of features describing the tissue. In tissue with layered structures, the reflectivity, spectral type, and distance of boundaries are determined.

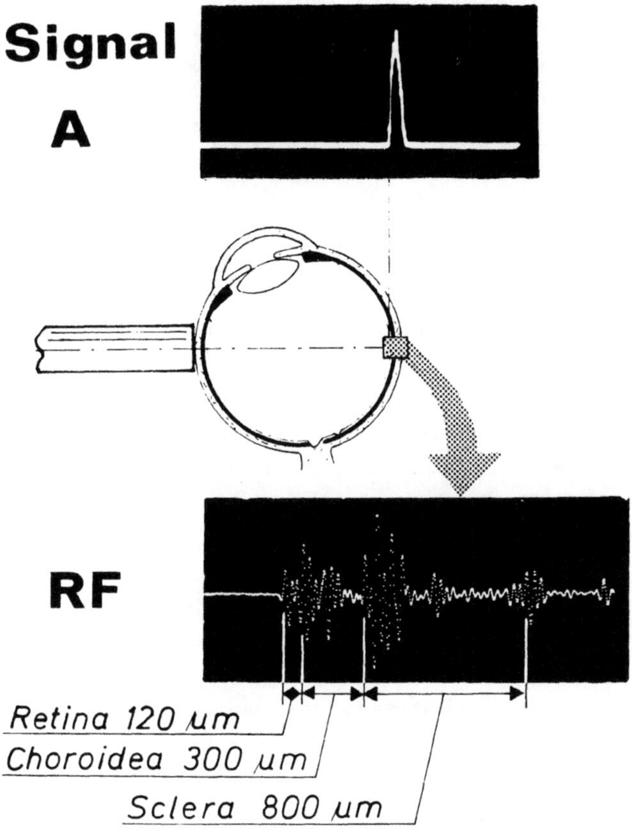

*Figure 3. The ultrasonic evaluation of the wall of the eye ball. Top: conventional A-mode signal. No details on tissue structure.
Bottom: the high resolution Radio-Frequency signal (RF); it reveals the layers of retina, choroid and sclera, and further structural information (schematically).*

For layer thickness measurement, cross correlation and spectral distribution proved suitable down to layers of 60 µm.
Tissues producing stochastic scattering ensembles (like tumours) are characterized by the distance, size, distribution, and spectral features of these ensembles. The ensembles are separated before the power spectra are derived. For characterization of tissue attenuation only these spectra are used, and are convolved with the power spectrum of the weighting signal.

3. In vivo-results.

The clinical evaluation of the system has shown:
A special potential of this technique is the characterization of thin tissue layers, i.e. the different layers of the eye wall, ranging between 0.08- 1mm (Fig. 3). Feature sets for the intact eye wall, the detached retina, and the "eye wall without retina" were compiled and demonstrated, including distance of boundaries (measurement capability down to 60 µ layer thickness), reflectivity, and spectral behaviour of the boundaries (Table I, Fig. 4). This serves for differentiation of vitroretinal diseases (direct and indirect approach for the detection of retinal detachment) and for tumour recognition by envelope layer analysis.
The evaluation of intraocular tumours (malignant melanoma of the choroid, metastasis), and tumour simulating lesions (subretinal blood, choroidal haemangioma, disciform macular degeneration, and others) made mainly use of a) this layer characterization for the "tumour surface", and b) of pattern analysis, and spectral attenuation measurements, performed at the internal tumour texture (see contribution K.M. Irion and D. Decker). These examinations are continued.
The authors expect other important clinical progress by combining tumour recognition with special A/B-mode scanning techniques, especially motional sensitive methods (A-, RF-, TM-, Doppler-methods) in tissue characterization. Measurement of blood flow-dependent tissue movements down to 10 µm were achieved with the above RF-system.
For these applications overlapping with "microbiometry" additional electronic counting systems for evaluation of the RF-signals are under study.

Table I. tissue features for the recognition of detached retina and vitreous membrane; examples of 4 clinical cases :
Upper part: direct diagnostic approach.
left: detached retina
right: vitreous membrane
Lower part: indirect diagnostic approach.
left: back wall without retina
right: normal back wall
(without scleral features)

FEATURES	DETACHED RETINA		VITREOUS MEMBRANE	
Boundary surfaces	I	II	I	II
Layer thickness/μm	128 const		120 ... 700 var	
Centre frequency/MHz	10.5	14.5	15.5..17.5	17.5
Spectral type	SP	SP	ST	ST,FS
Additional layer Δ s/μm	-	95 var	120 ... 200	var -
Reflectivity	- 28		- 32	

	BACK WALL WITHOUT RETINA			NORMAL BACK WALL (RETINA and CHOROID)		
Boundary surfaces	I	II	III	I	II	III
Layer thickness/μm		150..240	240..440 const		131 const	542 const
Additional layer Δ s/μm	-	90..220 var	-	-	121 const	-
Centre frequency/MHz	-	8.0..13.5	8.5..16.0	12.6	14.5	14.0
Spectral type	FS	SP	SP	SP	SP	SP
Reflectivity/dB		- 30			- 20	

SPECTRAL TYPE OF SURFACES
SP: specular surface

ST: scattering surface

FS: structured surface

LAYER THICKNESS
const = constant per file and per patient (n files)
var = constant per file; variable per patient (n files)

REFLECTIVITY
dB-level referenced to a flat target reflectivity (15 mV = 0 dB)

4. Application of ophthalmic results in other organs.

The above techniques are generally applicable in superficial small organs, allowing the use of high ultrasonic frequencies without heavy influence due to tissue layers covering over the region of interest. This is true for thyroid, testes, lymphnodes, skin. The thin layer characterization generally seems of interest for organs in which layered structures are present, like epithelium, endothelium, capsules, skin, muscle layers.

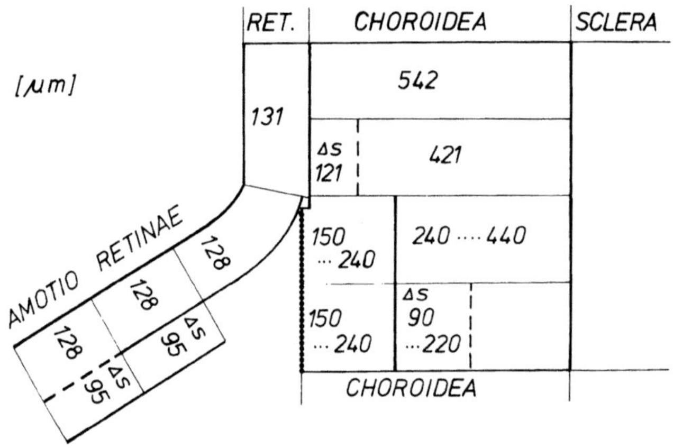

Figure 4. In vivo-results of computer-aided RF-signal analysis at the peripheral wall of the eye ball in 70 eyes.
Top : normal eye wall.
Bottom : left side: detached retina (Amotio retinae)
right side: eye wall without retina.
In every condition different biological types were observed. The scheme shows the layer thickness (μm) and the spectral type of the boundary surfaces (average values).

Spectral type of boundary surface :
Continuous lines = specularly reflecting surface
Dotted lines = structured surfaces.

REFERENCES

Trier, H.G. and Reuter, R. (1973): Eine Anlage zur halbautomatischen Klassierung verschiedener Formen von Gewebsechogrammen in Zeit-Amplituden-Darstellung. Proc. SIDUO IV, Paris 1971; M. Massin and J. Poujol, Eds., pp. 87-91 (Centre National d'Ophthalmologie des Quinze-Vingts, Paris).

Trier, H.G. and Reuter, R. (1973): Digital computer analysis of time-amplitude ultrasonograms from the human eye. I. Signal acquisition. J. Clin. Ultrasound 1, 150-154.

Trier, H.G. (1977) : Gewebsdifferenzierung mit Ultraschall. Medizinische und technische Grundlagen für eine erweiterte Gewebsdifferenzierung mittels Zeit-Amplituden-Signal im Impuls-Echo-Verfahren. Bibl. ophthal. No. 86 (Karger, Basel).

Trier, H.G., Lepper, R-D, and Reuter, R. (1979): Das Projekt "Rechnergestützte Gewebsdifferenzierung" der Arbeitsgemeinschaft Bonn/Stuttgart. I. Stand der Methodik in vivo. Proc. SIDUO VII, Münster 1978, in: Diagnostica Ultrasonica in Ophthalmologia, H. Gernet, Ed., pp. 35-39 (Remy, Münster).

Lepper, R-D, Reuter, R. and Trier, H.G. (1978): Digitization of high-frequency ultrasonic signals for tissue differentiation in ophthalmology. Biomed. Technik 23, 75-78.

Trier, H.G. (1980): Ultrasonic tissue characterization in the eye and orbit. In: Ultrasonic tissue characterization: clinical achievements and technological potentials. J.M. Thijssen, Ed., pp. 45-67 (Stafleu's Scientific Publ. Comp., Alphen a/d Rijn, The Netherlands).

Decker, D. and Irion, K.M. (1980): A-mode RF-signal analysis (frequency domain). In: Ultrasonic tissue characterization: clinical achievements and technological potentials. J.M. Thijssen, Ed., pp. 231-242 (Stafleu's Scientific Publ. Comp., Alphen a/d Rijn, The Netherlands).

Decker, D., Trier, H.G., Irion, K.M., Lepper, R-D. and Reuter, R. (1980): Rechnergestützte Gewebsdifferenzierung am Auge. Ultraschall in der Medizin 1, 284-296.

Trier, H.G., Decker, D., Müller-Breitenkamp, R. Irion, K.M. and Otto, K.J. (1981): The recognition of detached retina and vitreous membranes by means of radio frequency signal analysis. Proc. SIDUO VIII, Nijmegen 1980. In: Ultrasonography in Ophthalmology, J.M. Thijssen and A.M. Verbeek, Eds., pp. 419-430 (W. Junk Publ., The Hague, Boston, London).

Weigelin, E. and Trier, H.G. (1981) : Méthodes objectives de caractérisation tissulaire aux ultrasons: prospectives d'application clinique en ophthalmologie. 87. Congres Soc. franç. Ophthal., Paris, May 1981 (in press).

Trier, H.G. (1981): Rechnergestützte Ultraschalldiagnostik am Auge in Forschung und klinischer Anwendung. Dreiländertreffen Ultraschalldiagnostik 81, Graz Oktober 1981 (in press).

Lepper, R-D., Trier, H.G. and Reuter, R. (1981): Ultrasonic measurements at the posterior wall of living human eyes. Ophthal. Res. 13, 1-11.

Discussion

H.G. TRIER and R.D. LEPPER

Tissue characterization in Ophthalmology

Question by: I. Bredael
How do you cancel out the time-jitter of the acquisition system (about ca. 2.5 µs).

Reply by: R.D. Lepper
We do not because we do not need to!
With the Transient Digitizer R 7912 averaging in the time domain is quite simple because that instrument practically has no time-jitter due to the more analogeous structure of it.
With the new 7612 AD we only take single-sweep-signals and we do not average in the time domain. If the need for that arises we simply have to synchronize the A/D convertor and the transmitter by one master clock.

Question by: L. Masotti
Had you drift problems to solve in using Tektronix 7912 equipment ?

Reply by: R.D. Lepper
Concerning the long term stability: yes, with respect to the daily routine: no.

Question by: C.R. Hill
You mentioned in your talk that you have experimented with the use of Dr. Nicholas' diffraction technique for application in ophthalmology. Can you say wether it seemed to be of value?

Reply by: H.G. Trier
We examined in vitro thyroidal gland diseases, using transducers of our ophthalmology applications ($f_c \simeq 14$ MHz). From the broadband signal, up to now the 10-14 MHz frequency band, visual evaluation of the patterns was not successful. Perodicity analysis of the patterns together with amplitude evaluation seems however promising.
The method was tested with a tissue model with isotropic distribution of scatterers with 7 µm diameter with tissue-like impedance matching.

Further discussions and remarks

One of the important problems in ophthalmology is the differentiation of a subretinal haemorrhage from (malignant) subretinal tumours. This problem has met much attention till now and the ultrasonic differentiation, either with the video echograms, or by using the RF signals, while using computer analysis of various kinds seems very rewarding. Extrapolation of the knowledge and experience gained in Ophthalmology to the diagnosis of diseases of the thyroid and of lymphnodes may be feasible. Furthermore, the membrane characterization techniques devised in Ophthalmology could be tried in other organs, like stomach wall, gut walls and vessel walls. Some technical problems connected to the rescaling of the frequency range, the effects of intervening tissues etc. should be considered, but it seems worthwhile to make a try!

SYNTHETIC FOCUSSING IN MEDICAL ULTRASOUND

A. J. Berkhout, P. R. Mesdag, D. de Vries and P. van Riel
Delft University of Technology, Dept. of Applied Physics, Group of Acoustics,
Delft, The Netherlands

ABSTRACT

Presently, puls-echo techniques present the main activity in the acoustics group of the Delft University of Technology.
Applications lie in Geophysics, nondestructive testing and medical imaging.
The philosophy of the group is based on two principles :

a. Incorporation of wave theory (according to the acoustic wave equation) for inhomogeneous media in the physical model of acoustic echo data.
b. Use of linear inverse theory in the digital analysis and processing techniques of RF echo measurements.

With respect to the medical applications, the following projects are of interest :

1. Improvement of spatial resolution by means of computerized acoustic focussing, using wave field extrapolation techniques.
2. Tissue identification by means of a second-order statistical investigation of focussed data.
3. Velocity estimation with the aid of coherency methods and minimum-entropy techniques.
4. Model verification by means of parameter estimation.

I. SYNTHETIC FOCUSSING

The methods in this chapter are based on a computerized acoustic focussing technique, named "synthetic focussing". This technique was described earlier by Berkhout, Ridder and v.d. Wal (1, 2, 3). Synthetic focussing can best be described as a two-dimensional (or three-dimensional) inverse-filter technique, by means of which distorting propagation- and transducer characteristics may be optimally compensated for.

Basically synthetic focussing consits of two steps : inverse wave-field extrapolation and imaging.
In Figure 1 an acoustic measuring system is divided into various sub-systems, each of which can be described mathematically as a separate matrix (Berkhout (4)). For certain source-detector combinations these matrices may be inverted (Berkhout (1)), such that the measured pressure distribution on surface z_o, $p(x,z=z_o,t)$, will be transformed into pressure distribution $p(x,z=z_m,t)$ as it would be measured on surface z_m inside the medium. This part of synthetic focussing is called inverse wave-field extrapolation, as the wave-field is mathematically extrapolated back from surface $z=z_o$ to surface $z=z_m$. Proper inverse wave-field extrapolation from z_o to z_m removes all propagation effects of the medium between z_o and z_m.

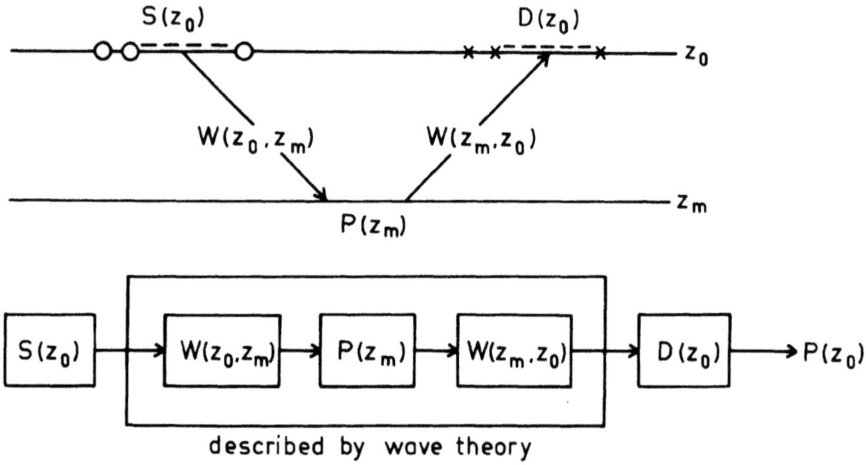

Figure I-1: Division of a measurement system into various sub-systems (from: Berkhout (4)).

The combination of sources and detectors (or: data acquisition system) determines the matrix configuration. A particular choice depends on the practical measuring situation and on which physical parameters one wishes to measure (Ridder, Berkhout and v.d. Wal (2,3)).
After the wave-field has been extrapolated back to surface z_m we may use the t=o registration as an estimate for the local reflectivity distribution. This is the imaging part of the synthetic focussing technique. The final product of the synthetic focussing technique is a pressure distribution p(x,z,t=o), which may be used as an optimal estimate for the reflectivity distribution throughout the entire image.
If we now visualize the synthetic focussing technique as one step, we see that we are transforming the measured pressure distribution p(x,z=o,t) into p(x,z,t=o), which is no more than a coordinate transformation from the coordinate t to the coordinate z. In the next paper the synthetic focussing technique will be used as a tool, so we will only concern ourselves with the measured pressure distribution p(x,z=o,t) and with the synthetically focussed pressure distribution p(x,z,t=o).

REFERENCES

1. Berkhout, A.J. (1980): Seismic migration, imaging of acoustic energy by wave field extrapolation. Elsevier Scientific Publ. Company, Amsterdam.
2. Berkhout, A.J., Ridder, J. and van der Wal, L.F. (1982): Acoustic imaging by wave field extrapolation, Part 1: Theoretical considerations. In: Acoustical Imaging, Vol. 10, pp. 513-567; Eds. P. Alais and A. Metherell, Plenum, New York.
3. Ridder, J., Berkhout, A.J. and van der Wal, L.F. (1982): Acoustic imaging by wave field extrapolation, Part 2: Practical aspects. In: Acoustical Imaging, Vol. 10, pp. 513-567; Eds. P. Alais and A. Metherell, Plenum, New York.
4. Berkhout, A.J. (1981): Wave field extrapolation techniques in seismic migration, a tutorial. Geophysics, Vol. 46, Nr. 12, pp. 1638-1656.

II. ON THE RELATION BETWEEN DIFFRACTION PATTERNS AND AN OPTIMALLY FOCUSSED IMAGE

Until recently tissue characterization has known two different approaches. One approach starts from an image as produced by commercially available ultrasound machines. From this image one characterizes tissue by its contours and by local reflectivity parameters.

The other approach starts from a sample volume, building-up diffraction patterns and trying to characterize tissue locally from these diffraction patterns.

In this chapter we would like to put forward, that both approaches to tissue characterization are theoretically exactly the same and that one approach will principally give no more information than the other. This chapter is based on the synthetic focussing technique, and the output of this technique will be discussed here.

First we will characterize the system reponse of our imaging system. For this we will use a model of one point diffractor, as this may be seen as a spatial reflectivity delta-pulse. Analoguous to other disciplines the system response can best be studied in Fourier space. In our case of a two-dimensional image this will be the two-dimensional Fourier space.

Figure 1 shows the two-dimensional Fourier transform of the synthetically focussed diffractor. If our imaging system would have had no limitations, then we would see a white spectrum in both lateral and axial sense.

With Figure 1 we have characterized the acoustic spectral window, through which we are able to look at the point diffractor, and outside of which no information can be gathered after data acquisition. Next we shall look at two point diffractors.

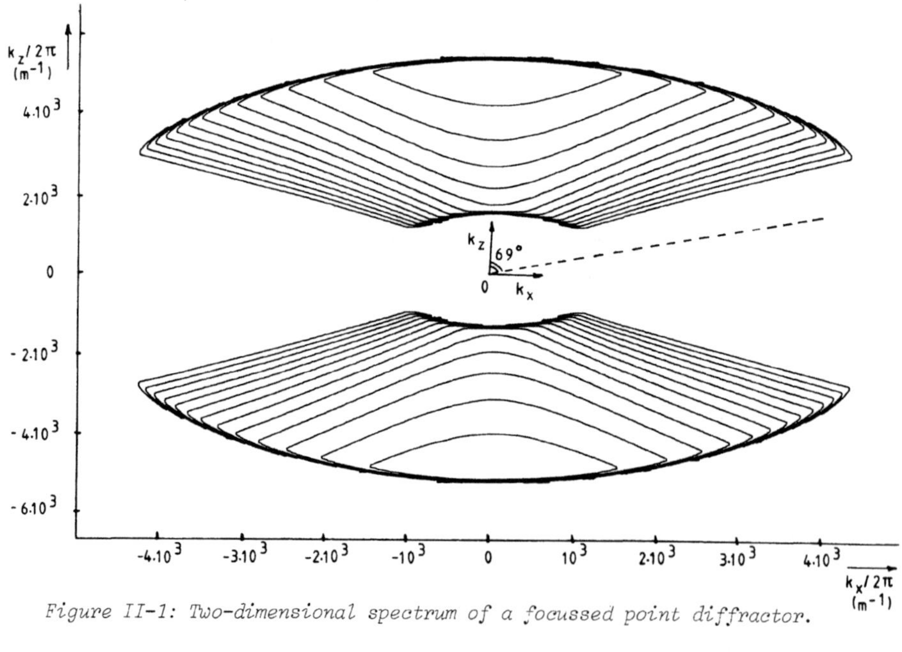

Figure II-1: Two-dimensional spectrum of a focussed point diffractor.

The spatial reflectivity distribution of two point diffractors may be characterized as:

$$r(x,z) = \{\delta(x-x_o) + \delta(x-x_1)\}\ \delta(z-z_o),$$

of which the modulus of the two-dimensional Fourier transform is

$$|\tilde{r}(k_x,k_z)| = 2\cos\ \{\tfrac{1}{2}k_x(x_1-x_o)\}. \quad \text{(Figure 2)}.$$

Like with the single point diffractor the focussed result is deformed by the imaging system, this may be seen as a two-dimensional convolution of our reflectivity distribution r(x,z) with the focussed result of the single point diffractor.

Consequently the two-dimensional Fourier fransform of the two focussed diffractors will be a multiplication of the Figures 1 and 2. This is shown in Figure 3.

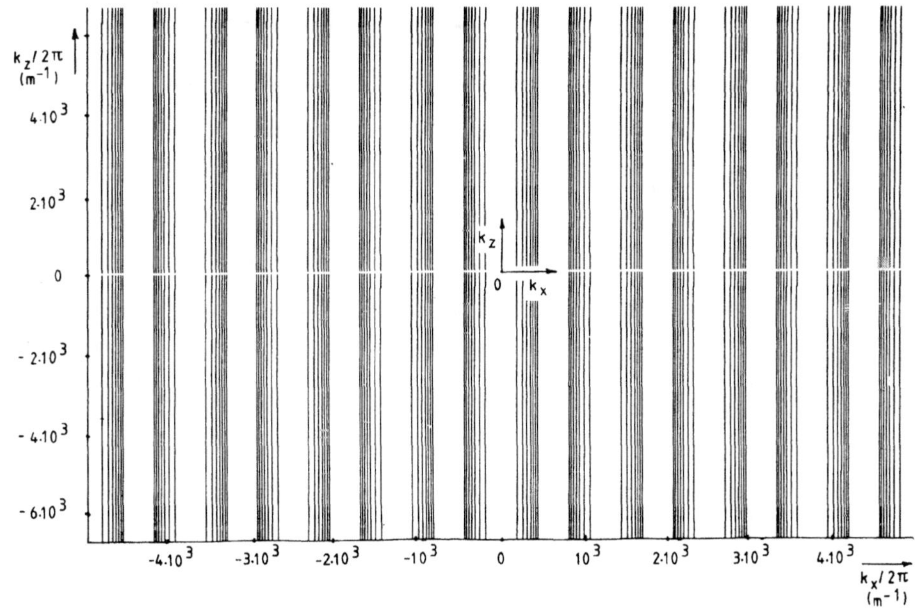

Figure II-2: Theoretical diffraction pattern from two point diffractors.

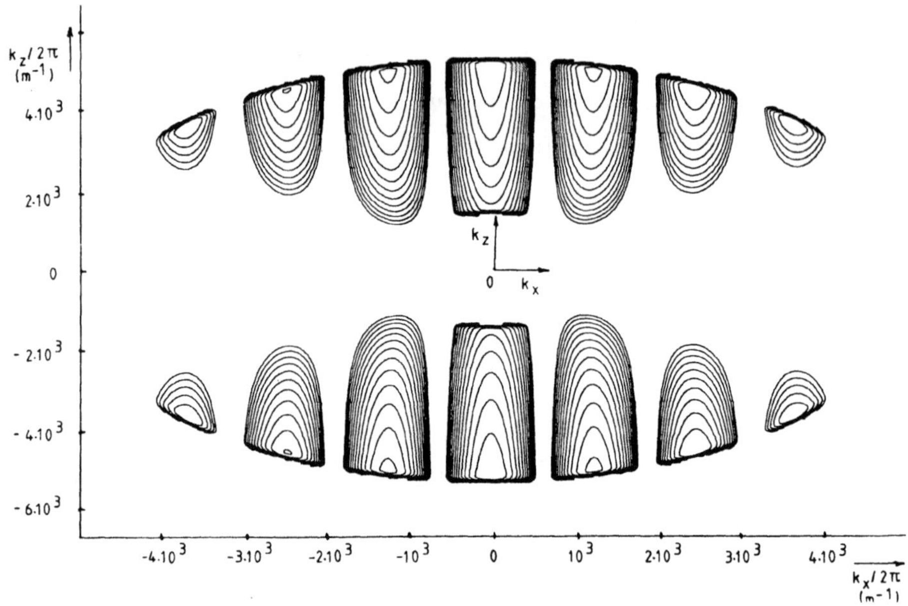

Figure II-3: Measured diffraction pattern from two point diffractors.

According to amongst others Waag (1) and Nicholas (2) there exists a Fourier relationship between the spatial distribution of discontinuities and the backscattered diffraction pattern therefrom. It can be shown that the mentioned diffraction patterns are subsets of our two-dimensional Fourier space. Before tissue characterization by means of diffraction patterns is possible, two conditions must be met :

Firstly far field conditions must be present to ensure the Fourier relationship (Waag (1)). These far field conditions are synthetically generated by our two-dimensional Fourier transform.

Secondly high resolution must be present to be sure we are looking at the tissue in our sample volume only. Through synthetic focussing this high resolution is present over the entire image.

REFERENCES

1. Waag, R.C. (1980): Theory and measurements of ultrasonic scattering for tissue characterization. In: Acoustic Imaging, Vol. 9, Ed. Wang, Plenum Press, New York.
2. Nicholas, D. (1979): Ultrasonic diffraction analysis in the investigation of liver disease. British J. Radiol., 52, pp. 949-961.
3. Mesdag, P.R., Berkhout, A.J. and Ridder, J. (1982): A wave theoretical approach to tissue characterization. Submitted for publication in J. Acoust. Soc. Amer.

III. VELOCITY DETERMINATION

In synthetic focussing procedures, the sound propagation velocity value of the medium between aperture and target has to be inserted as a parameter into the extrapolation operator. Errors in this value may highly affect the lateral focussing quality of the extrapolated result, as is shown in Figure III-1, where the focussed pattern of a harmonic dipole scatterer is displayed for aperture angles of 15^{o}(a-e) and 60^{o}(f-j). In this example, the vertical travel time for central aperture coordinate x=o has been exactly corrected in all cases; Figure III-1, c and h give the (optimal) focussed result for correct extrapolation velocity insertion, whereas the other figures give the extrapolated patterns for velocity values deviating -20%, -10%, +10% and +20% from the true propagation velocity. It is seen that especially for large aperture angles velocity errors strongly deteriorate the focussing quality of the extrapolated pattern. Therefore, the medium velocity should be determined as exactly as possible from the registered data before starting the focussing procedure, particularly if large apertures are used.

For this purpose, several techniques are being developed and improved in the framework of our velocity determination project.

* A velocity estimation method for layered media, well-known in the exploration seismology world as the "Dix-method", estimating the velocity value from coherence-criteria applied to Common Midpoint (CMP)-registered data, has been improved by introducing extrapolation- and "velocity-replacement"-procedures. The method now can be succesfully applied to more-layer structures with non-parallel or even curved boundary configurations.

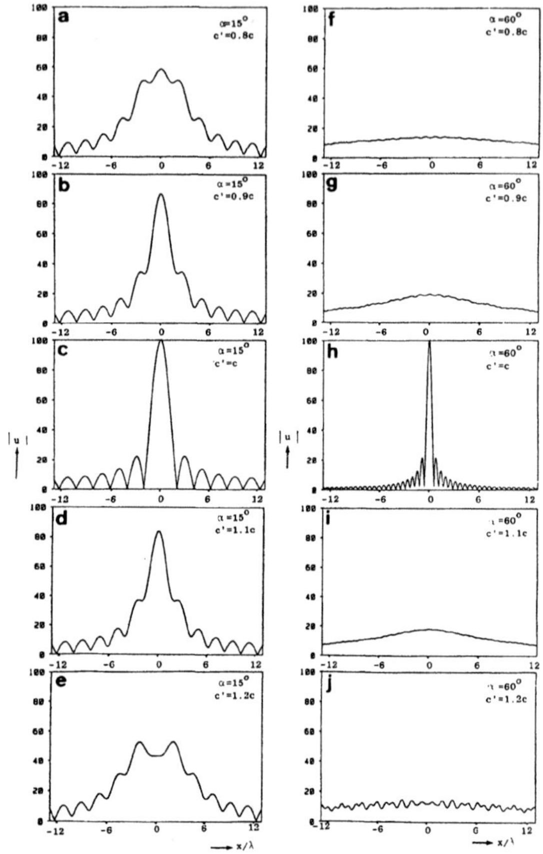

Figure III-1: Focussed pattern of a harmonic dipole scatterer.

Figs. III-1, f-j above suggest that for large aperture angles, the height and/or widths of the focussed pattern may be used as a measure for the accuracy of the extrapolation velocity used. Possibly, for these angles height- and pattern width criteria could form a simpler alternative for the coherence criterion described above; investigations on this point have been started.

* If the medium does not have a layered structure, but contains some distribution of scatterers, as in many medical applications, coherence criteria are not suitable. In such cases, application of minimum entropy criteria could solve the problem. Research on the possibilities of such techniques have recently started.

IV. MODEL VERIFICATION BY MEANS OF PARAMETER ESTIMATION

In the inversion of echo-acoustic data spatial and axial deconvolution may be considered separately. In this introduction some aspects of axial deconvolution are discussed.
The basics of axial deconvolution can all be derived within the framework of the simple convolutional model describing the forward problem :

$$s(t) = w(t) * r(t) + n(t)$$

where s(t) is the reponse, w(t) is the wavelet, r(t) is the reflectivity function and n(t) is additive noise. The wavelet w(t) is in all echo-acoustical applications bandlimited. The reflectivity function is in general adequately modelled by :

$$r(t) = \sum_{i=1}^{N} a_i.\delta(t-\tau_i)$$

i.e., a series of N discrete reflectors with amplitudes a_i and lag times τ_i.$\delta(.)$ is the Dirac impulse function. An example of the synthetic construction of the response function is given in Figure IV-1, a-c.
The goal of axial deconvolution is to estimate the reflectivity parameters by removing as much as possible of the effect of the wavelet without unacceptably enhancing the noise component.

Conventionally deconvolution is approached through linear inverse filtering. This will in general result in a sharpening of the response function. Inherent to the method is however that the resolution attainable is

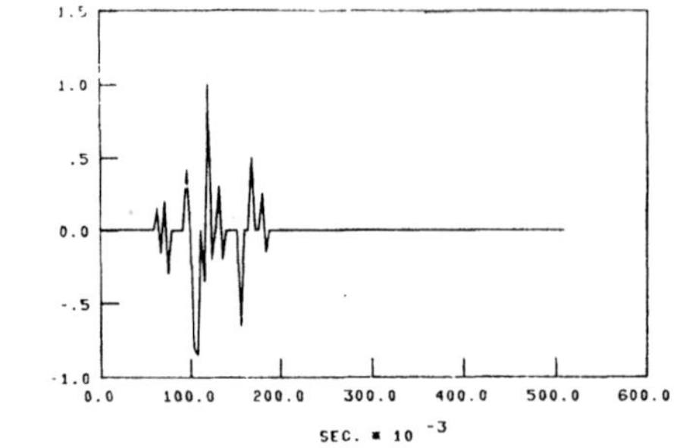

Fig. IV-1a: The actual reflectivity

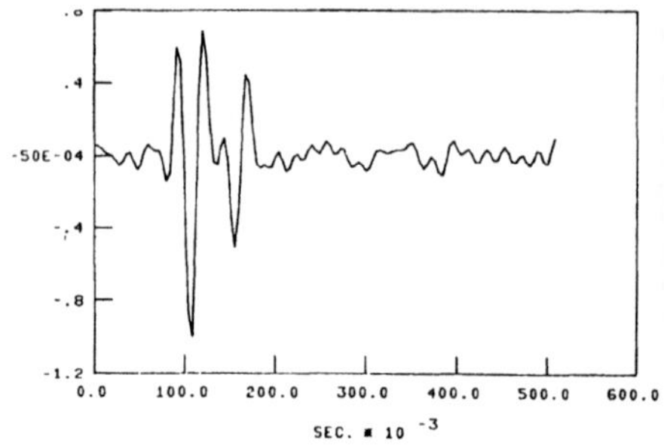

Fig. IV-2: The estimate with inve filtering

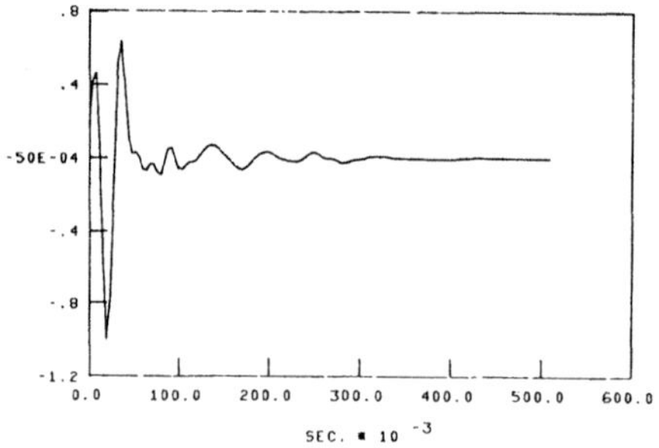

Fig. IV-1b: The wavelet

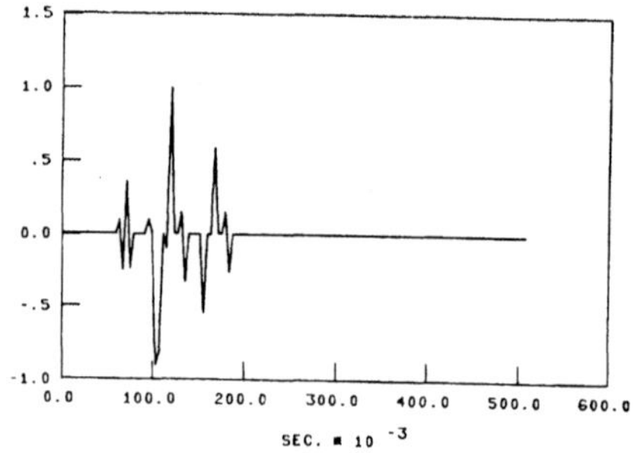

Fig. IV-3a: The initial model

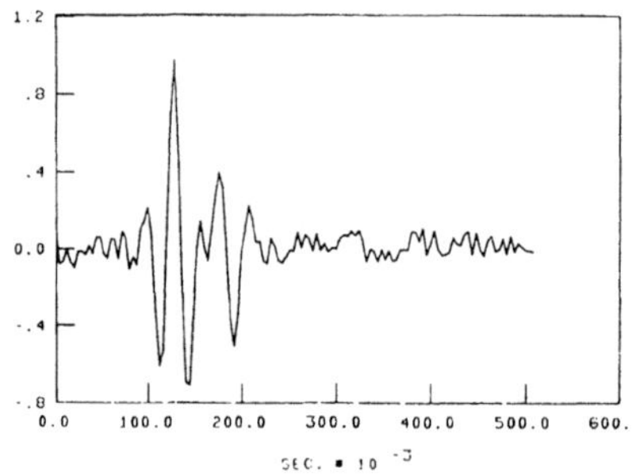

Fig. IV-1c: The response

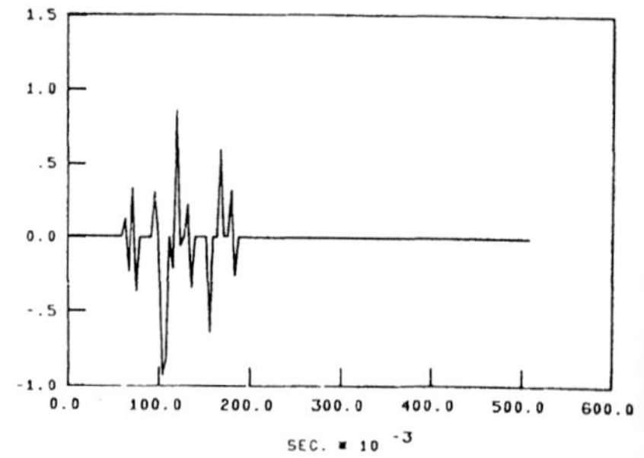

Fig. IV-3b: The estimated model

limited by the available data bandwidth. In Figure IV-2 the results of optimal inverse filtering of the data shown in Figure IV-1c are given. Notice that, due to the limited resolution, only the major reflectors have been resolved.

A theoretical investigation on the information retrieval from bandlimited data has shown that the volume of information retrievable is limited. In inverse filtering no assumptions, such as the number of reflectors, are made on the reflectivity function to be estimated. For this reason the results are bandlimited. Now, in many applications, e.g. Ophthalmology, it is known a priori that the number of parameters to be estimated from the data is limited. These can therefore in principle be recovered with a resolution higher than follows from the available data bandwidth.

The method through which prior information is most readily incorporated is modelling. The determination of the model parameters is direct and can for instance be accomplished by parameter estimation techniques.
If parameter estimation procedures are to be applied, the forward problem first has to be reformulated in matrix terms. For the problem at hand this results in a convolution matrix linking the parameter space to the data space.
For bandlimited data, as discussed, the amount of information which can be retrieved is limited. The implication is that the forward system may be ill conditioned leading to various difficulties in solving the inverse problem. The estimation or inversion method which most effectively counters these problems is based on the stochastic reformulation of the forward problem combined with the Singular Value Decomposition (SVD) or diagonalization of the forward operator. In this framework the size of the diagonal elements is related to the assumed noise levels in the stochastic model. From this analysis an inverse operator is derived which inverts only for those data and parameter combinations characterized by S/N>1. Applied to the data, this inverse operator generates those estimates with the highest resolution, i.e. highest information content, compatible with the noise levels.
In Figure IV-3a-b, results are shown for the linear estimation problem obtained by assuming that the lag times τ_i are known a priori. Figure IV-3a shows an initial model with 20% amplitude errors compared to the actual model shown in Figure IV-1a. Figure IV-3b shows the updated model obtained with the estimation procedure described. The amplitude errors have been reduced to 5%.

The proposed method has large potential; it is a generalization of conventional least squares parameter estimation techniques, it produces results ranging from full resolution for sparse models to the resolution following from data bandlimits for complex models, it allows the incorporation of initial information in various forms to enhance resolution, it is very robust with respect to noise, the method is applicable to a large range of systems including continuous time parameters and time variant operators.
Success in application is however not guaranteed. Most critical is that the method is strongly dependent on the accuracy of the initial models. Secondly, limited resolution introduces a measure of ambiguity, the results are partly non-unique. Thirdly, the estimation of non-linear parameters involves iteration and can only lead to local optima. Lastly, the forward operator has to be known with reasonable accuracy.

REFERENCES

Parker, R.L. (1977): Understanding inverse theory. Arr. Rev. Planet. Sci. Vol. 5, pp. 35-64.

Riel van, P. and Berkhout, A.J. (1982): The limits of the modelling of the seismic reflectivity function in terms of data bandwidth. (To be published).

The General Linear Inverse Problem: Implications of Surface Waves and Free Oscillations for Earth Structure. Rev. Geophys. Space Phys., Vol. 10, pp. 251-285.

Discussion

A.J. BERKHOUT and P.R. MESDAG

Synthetic focusing in medical ultrasound

Question by: P. Lewin

Would you please comment on the processing time of the synthetic focusing method?

Also, could you tell us what was the frequency of your experiment?

Reply by: L.F. v.d. Wal

In the final synthetic focusing system, processing time on a 256 x 2048 data set, will be less than 60 seconds.

The spectral bandwidth in our simulations and practical experiments was ca. 2-4 MHz.

Question by: L. Masotti, G. Castellini

How did you solve, in synthetic focusing, the problems deriving from the presence of grating sidelobes due to the value of the interspace between the centers of two adjacent elements and to the directivity function of the single element?

Reply by: L.F. v.d. Wal

The problem of grating lobes is acknowledged by both speakers. Using Zero Offset Data Acquisition, spatial sampling should take place at intervals $\Delta x < \lambda/2$ to avoid spatial aliasing. Due to practical limitations in transducer manufacture spatial sampling takes place at intervals $\Delta x = 0.4$mm ($\approx\lambda$, for f_{max}). So grating lobes may be present in the focused image. The finite size of a single transducer element (0.3 mm) will however act as a spatial filter, adequately suppressing echo-signals coming in from grating lobe positions.

Research work to avoid spatial aliasing, using the linear array described above, is in progress using different types of data acquisition.

Question by: J.M. Thijssen
If I am well, the velocity estimate should be accurate in order to obtain proper focusing. Are you with the minimum entropy using a multilayered medium model or do you have to restrict to a single velocity and how do you conceive this method in medical diagnosis?

Reply by: L.F. v.d. Wal
In principle the minimum entropy method may be used in any medium: "multi-layered" or a medium formed by a distribution of scatterers. The method may be used to determine either the best fit for the overall propagation velocity (bulk velicity) or be applied to sub-area's in the focussed image to estimate the local velocities, which then may be optimized independently in each area.

Further discussions and remarks

The feasibility of the synthetic focussing in medical diagnosis is highly dependent on the speed of production of pictures. By making use of hardware additions to the digital computer (array processor) and improving the data transport rate a calculation time reduction from many hours to minutes has been achieved.
The problem of a limited aperture transducer was considered. Implications for the axial as compared to lateral resolution achieved and for the optimization of imaging as compared to diffraction analysis were discussed, and the question what synthetic focussing might bring for the analysis of the fine "speckling" in the images of solid tissue was risen.
Finally the general problem that echographic imaging produces a two dimensional axial tomogram and, therefore, the influence of the information in the third dimension is neglected, was discussed. No easy solution was found, may be with the exception of future ultrasonic CT scanner employing a waterbath.

STATUS OF TISSUE CHARACTERIZATION WORK AT THE INSTITUTE OF CANCER RESEARCH AND ROYAL MARSDEN HOSPITAL

C. R. Hill
Institute of Cancer Research, Royal Marsden Hospital, Sutton, Surrey, England

INTRODUCTION

The work of our research group in this field commenced in 1969 and has been concerned generally with development of methods for cancer investigation. Here conventional pulse-echo imaging has a central role and part of our objective has been to understand and quantify the ultrasound-tissue interactions on which imaging depends, and thus to work towards improvements in the "diagnostic" quality of images. Coupled with this we have tried to develop ideas for objective classification of tissue type in a manner that would complement conventional imaging methods.

Detailed results of this work have been reported in a large number of publications and it will only be possible to review these here in outline. Aspects of the earlier phases of our work have been described in previous synopses (e.g. refs 13,14,17) and have also been discussed in the context of related work in the field in two recent review chapters (15, 16).

For the purpose of the present report it will be convenient to consider the work of our group under the headings : measurement of tissue properties, applied diffraction analysis, B-scan image analysis and image enhancement, and blood flow and tissue movement studies. This report will concentrate on technical aspects of the work but it should be borne in mind that our group is jointly clinical and scientific/technical, and that a high proportion of day-to-day practical "tissue characterization" is carried out by the eye and brain of diagnosticians working directly from B-scan images.

Measurement of Tissue Properties

From its commencement, the direction of our work has been coloured by the belief that the class of ultrasound/tissue interaction that is likely to be most informative diagnostically is that of scattering. However in attempting, in 1969, to scan the available literature we were able to find only two references to acoustic scattering by human or animal tissues (8) and, as a consequence, it was necessary to carry out considerable ground work in technique and methodology in addition to carrying out experimental studies of various aspects of scattering behaviour. Scattering measurements cannot be made adequately without good knowledge of attenuation coefficient, and preferably also sound speed, in the tissue or material under study. For this reason, as well as for their intrinsic interest, we have also carried out sets of measurements of these two quantities. In almost all cases our approach has been to attempt to measure these quantities as continuous functions of frequency over a considerable bandwidth (generally using two or three medium-to-broad band transducers, see, e.g. refs 10,11).

A set of "housekeeping" investigations has been carried out on the effects of various conditions of measurement on the values observed for various acoustic properties of tissues, in an attempt to establish the relevance to in vivo behaviour of laboratory measurements carried out on excised tissues. These investigations have included effects of time (and storage conditions) between death (or excision) and measurement (2), effects of the various major classes of tissue fixative (4) and dependence on measurement temperature (3). It has been shown in this work for example that storage at 20°C for 48 hours leads to appreciable changes in tissue scattering behaviour but not in attenuation coefficient, and that fixation using 4% formaldehyde leads to only rather small, although measurable, changes in these properties. One particular gap in our knowledge remains here: that of the changes, if any, in acoustic properties of tissues that may occur immediately on death or excision (e.g. as a result of cessation of blood supply).

Measurements have been made and reported of various acoustic properties of a number of human, and some animal tissues. Particular tissue types that have been studied here include normal liver, metastatic tumours in liver, liver modified by various diffuse pathologies, spleen, brain and skeletal muscle. Properties that have been measured are : attenuation coefficient, speed of sound, backscattering cros-section (30,27) and, more recently (Nassiri, to be published) total and differential (angular) scattering crossections. An example of a particular study here is that of the relationships of both liver pathology and tissue composition to attenuation coefficient and speed of sound : the latter quantity for example is consistently lower for tumour than for surrounding normal liver, with the difference being due primarily to the higher water content of the tumour (5,6). In another study (21) attenuation coefficients of skeletal muscle were found to be about 50% greater for propagation along, as compared with normal to the pulse orientation : an observation that poses the question of the mechanism of the excess attenuation.

Much interest has centred on diffractive scattering behaviour (phenomena that arise as a result of scattering by structures of the order of size of an acoustic wavelength). Initial work attempted to study this behaviour by measurements of frequency dependence (9) but attention subsequently shifted to measurements of backscattering amplitude at a single frequency as a function of relative orientation between beam direction and the scattering structure. Here we were able to show empirically that the diffraction patterns produced contained information, which could be extracted in objective, quantitative form, that would enable effective differentiation between certain groups of tissue types in vitro (28). These results have focussed our attention on investigation, by theoretical and experimental methods, of the fine-scale acoustic structure of tissues that, presumably, is responsible for this specific diffraction behaviour. To this end we have carried out considerable work on theoretical modelling of tissues (22, 26,1). We are collaborating with the group at University College London on acoustic microscopic studies of tissues - aimed at deriving information on spatial variations of density and compressibility (31) and are attempting to complement such information from experimental studies of the angular scattering behaviour of tissues of clearly differing structure, particularly blood, liver and skeletal muscle (Nassiri, in preparation).

Applied Diffraction Analysis

Arising out of some of our earlier laboratory work described above we have implemented in clinically applicable form an orientation diffraction analysis technique (18) with the objective of testing its potential for *in vivo* tissue differentiation. This has now been applied in preliminary trials in our Hospital in two series of patients with known or suspected disease of liver and thyroid respectively. In both cases (23,19) useful degrees of differentiation were demonstrated. In view of the prototype nature of the equipment this has been very encouraging and we have more recently been extending the work in two directions: (a) to incorporate the technique as part of a multi-element array transducer system, thus at the same time integrating imaging and analysis, and enabling acquisition of data to be sufficiently fast as to greatly reduce problems of movement artefact (29), and (b) to extend the scan geometry from a planar arc to a full solid angle, thereby much increasing the quantity of data collected and also opening up the possibility of characterizing tissues by some measure of any anisotropy that they may exhibit (Nicholas and Nicholas, in preparation).

Image Analysis and Enhancement

Hitherto very little work on diffraction analysis as such has been reported from outside our group. A related manifestation of the phenomena that underlie the method is in the fine spatial structure of conventional A- and B-scans, and this is a topic that has been taken up by a number of groups. Our work here started with that of Milan (32) in echo amplitude histogram analysis but the more recent phase of the work has been more concerned with spatial properties of the data. Our basic approach here has commenced from the position that a B-scan is the basic stock-in-trade of the diagnostician and has thus led us to apply statistical type analyses to the information contained in such B-scans. This involved a deliberate decision to ignore, for the time being, any additional information which may be coded as phase in the r.f. echo signal and which is then lost in the detection process that leads to the "video" signal. It should be noted however that we avoid in this process any unnecessary degree of frequency filtering or non-linear recording (loss of dynamic range), both of which can be expected to result in loss of information. We are now running clinical trials, involving liver disease, based on the use of a number of trial discriminators which can be derived from B-scan data stored in this way. Initial results here are again encouraging, and indicate a potential for useful differentiation of various disease conditions (24). We are also developing approaches to both characterizing and enhancing shape of discreet targets (e.g. blood vessels, tumours) which may be contained in a B-scan.

Thus, as far as concerns clinical trials of tissue characterization, particularly in oncology, our group can claim some reasonable experience, with two distinct technical approaches. One clear lesson of this however is that, in any such work, there must inevitably be a considerable period of technical development and clinical "learning" between demonstration of clinical feasibility and routine implementation. This arises partly because techniques of this type are instrumentation dependent - the data that they produce are functions of instrument parameters as well as of histo-pathology - and thorough optimization of technique is thus desirable before settling on a stable equipment configuration that can be the basis for definitive clinical trial and acquisition of a complete "learning set" of correlations between acoustic characteristics and established disease conditions. Associated with this is the considerable problem of obtaining definitive and statistically sound follow-up for a number of diagnostic problems.

Blood Flow and Tissue Movement Studies

We have recently been giving increased attention to ultrasonic methods of characterizing the dynamics of tissues in relation to malignant disease. This work falls into two parts : (a) Doppler investigations of tumour blood flow, and (b) analysis of pulse-echo signals aimed at characterizing patterns of movement of organized tissues.

Our investigations of tumour blood flow have been stimulated by the desire of clinicians to find sensitive and rapidly acting indicators of the response of tumours to therapy: the initial interest being in response of non-resectable breast tumours to hormone therapy. We have therefore adopted the apparently well proven techniques developed by the Bristol group (7) and are applying these in longitudinal studies in selected patients who have presented in the Hospital Breast Clinic. Initial studies indicate that variations in the Doppler characteristics of tumours generally correspond to differences in tumour volume although, in response to hormonal influences, the blood flow variations may well precede the volume changes (20).

In our work on tissue movement characterization we have examined two somewhat different approaches: (a) correlation analysis of time-sequential, spatially invariant A-scans (12), and (b) examination of the time changes in orientation diffraction patterns (25). It appears that both of these approaches are capable of *in vivo* characterization of tissues on the basis of their rheological properties.

CONCLUSION

This review covers the work of a large number of people working in both fundamental and clinically applied areas, and with many lines of collaboration both within and outside our own Hospital and Institute. The work reviewed has been funded by grants from a number of sources, including Medical Research Council, the Cancer Research Campaign, Science and Engineering Research Council, the Royal Marsden Hospital and the EMI Company.

REFERENCES

1. BAMBER, J.C. and DICKINSON, R.J. (1980): Ultrasonic B-scanning: a computer simulation. Phys. Med. Biol., 25, 463.
2. BAMBER, J.C., FRY, M.J., HILL, C.R. and DUNN, F. (1977): Ultrasonic attenuation and backscattering by mammalian organs as a function of time after excision. Ultrasound in Med. and Biol., 3, 15.
3. BAMBER, J.C. and HILL, C.R. (1979): Ultrasonic attenuation and propagation speed in mammalian tissues as a function of temperature. Ultrasound in Med. and Biol., 5, 149.
4. BAMBER, J.C., HILL, C.R., KING, J.A. and DUNN, F. (1979): Ultrasonic propagation through fixed and unfixed tissues. Ultrasound in Med. and Biol., 5, 159.
5. BAMBER, J.C. and HILL, C.R. (1981): Acoustic properties of normal and cancerous human liver: I - dependence on pathological condition. Ultrasound in Med. and Biol., 7, 121.
6. BAMBER, J.C., HILL, C.R. and KING, J.A. (1981): Acoustic properties of normal and cancerous human liver: II - dependence on tissue structure. Ultrasound in Med. and Biol., 7, 135.
7. BURNS, P.N., HALLIWELL, M., WELLS, P.N.T. and WEBB, A.J. (1982): Ultrasonic Doppler studies of the breast. Ultrasound in Med. and Biol. (in press).
8. CHIVERS, R.C. (1973): The scattering of ultrasound by human tissues. PhD Thesis, University of London.
9. CHIVERS, R.C., HILL, C.R. and NICHOLAS, D. (1974): Frequency dependence of ultrasound backscattering crossections: an indicator of tissue structure characteristics. In: Ultrasonics in Medicine, pp. 300-303, Editors M. de Vlieger et al., Excerpta Medica, Amsterdam.
10. CHIVERS, R.C. and HILL, C.R. (1975): Ultrasonic attenuation in human tissue. Ultrasound in Med. and Biol., 2, 25.
11. CHIVERS, R.C. and HILL, C.R. (1975): A spectral approach to ultrasonic scattering from human tissue: methods, objectives and backscattering measurements. Phys. Med. Biol., 20, 799.
12. DICKINSON, R.J. and HILL, C.R. (1982): Measurement of soft tissue motion using correlation between A-scans. Ultrasound in Med. and Biol. (in press).

13. HILL, C.R. (1974): Interactions of ultrasound with tissues. In: Ultrasonics in Medicine, pp. 14-20, Editors M. de Vlieger et al., Excerpta Medica, Amsterdam.
14. HILL, C.R. (1976): Frequency and angular dependence of ultrasonic scattering from tissue. In: Ultrasonic Tissue Characterization, pp. 197-206, Editor M. Linzer, NBS Special Publication 453, U.S. Government Printing Office (Dept. of Commerce).
15. HILL, C.R. (1980): Tissue characterization. In: Progress in Medical Ultrasound Vol. 1, pp. 11-18, Editor A. Kurjak, Excerpta Medica, Amsterdam.
16. HILL, C.R. (1981): Tissue characterization. In: Progress in Medical Ultrasound Vol. 2, pp. 19-25, Editor A. Kurjak, Excerpta Medica, Amsterdam.
17. HILL, C.R., CHIVERS, R.C., HUGGINS, R.W. and NICHOLAS, D. (1978): Scattering of ultrasound by human tissue. In: Ultrasound, its Applications in Medicine and Biology, pp. 441-493, Editor F.J. Fry, Elsevier, Amsterdam.
18. HUGGINS, R.W. and PHELPS, J.V. (1976): Bragg diffraction scanner for ultrasonic tissue characterization in vivo. Ultrasound in Med. and Biol., 2, 271.
19. MERTON, J., NICHOLAS, D., HILL, C.R., GROVER, S., QUEENAN, M. and COSGROVE, D.O. (1982): Ultrasonic diffraction scanning of the thyroid. Ultrasound in Med. and Biol. (in press).
20. MINASIAN, H. and BAMBER, J.C. (1982): Study of mammary tumour biology using ultrasonic Doppler assessment of blood flow. Ultrasound in Med. and Biol. (in press).
21. NASSIRI, D.K., NICHOLAS, D. and HILL, C.R. (1979): Attenuation of ultrasound in skeletal muscle. Ultrasonics, 17, 230.
22. NICHOLAS, D. (1978): An introduction to the theory of acoustic scattering by biological tissues. In: Recent Advances in Ultrasound in Biomedicine, Chapter 1, Editor D.N. White, Research Studies Press.
23. NICHOLAS, D. (1979) Ultrasonic diffraction analysis in the investigation of liver disease. Brit. J. Radiol., 52, 949.
24. NICHOLAS, D. (1980): B-scan image quantitation of human hepatic tissue. In: Recent Advances in Ultrasound Diagnosis, Vol. 2, pp. 41-43, Editor A. Kurjak, Excerpta Medica, Amsterdam.
25. NICHOLAS, D. (1980): The analysis of soft tissue movements. In: Ultrasonic Tissue Characterization: Clinical Achievements and Technological Potentials, pp. 281-285, Editor J.M. Thijssen, Stafleu, Brussels.
26. NICHOLAS, D. (1982): Time-frequency domain analysis: one dimensional phantom studies. Phys. Med. Biol. (in press).
27. NICHOLAS, D. (1982): Evaluation of backscattering coefficients for excised human tissues: results, interpretation and associated measurements. Ultrasound in Med. and Biol. (in press).
28. NICHOLAS, D. and HILL, C.R. (1975): Acoustic Bragg diffraction from human tissues. Nature, 257, 305.

29. NICHOLAS, D., COSGROVE, D.O. and PUSSELL, S. (1980): Real-time diffraction scanning. Brit. J. Radiol., 53, 1025.

30. NICHOLAS, D., HILL, C.R. and NASSIRI, D.K. (1982): Evaluation of backscattering coefficients for excised human tissues: principles and techniques. Ultrasound in Med. and Biol. (in press).

31. SINCLAIR, D. (1981): Tissue and materials characterization using scanning acoustic microscopy. PhD Thesis, University of London.

32. TAYLOR, K.J.W. and MILAN, J. (1976): Differential diagnosis of chronic splenomegaly by grey scale ultrasonography: clinical observations and digital A-scan analysis. Brit. J. Radiol., 49, 519.

CLINICAL IMPLEMENTATION OF ULTRASONIC TISSUE CHARACTERIZATION AT THE ROYAL MARSDEN HOSPITAL

D. Nicholas
Institute of Cancer Research, Royal Marsden Hospital, Sutton, Surrey, England

INTRODUCTION

In its simplest form tissue characterization may be defined as the identification of one or more parameters relating to a region of tissue which are well correlated with the type or condition of that tissue. This implies that existing visual interpretation of ultrasonic images is also a form of "tissue characterization" albeit a subjective one. Although much work has been conducted in the laboratory using excised tissues or phantoms, only those techniques which have resulted in clinical trials will be discussed here. An accompanying report details the overall work of our research group.

For the purpose of the present report it will be convenient to discuss the different aspects of tissue characterization under the headings: diffraction information, tissue motion studies, and B-scan image analysis and image enhancement. These form the major categories into which our clinical trials are best placed yet many related topics also exist which, for brevity, will not be discussed herein.

Diffraction Information

Our objective has been to establish a method by which a quantitative relationship between the scattered ultrasound echoes and the structural composition of the interrogated tissues could be obtained. This, of course, demands that any pathological abnormality in a tissue must be related to a structural variance from normal: a fairly realistic premise. Currently available A-scan techniques (1-5) are basically limited to the assessment of tissue separations projected into the direction of the acoustic beam. Measurements performed in this manner will only accurately describe the true situation if the structures are either aligned normal to the A-scan direction (as in ophthalmology) or are isotropically distributed as a truly homogeneous medium. If neither of these conditions is satisfied then an assessment in two-dimensions will provide more accurate information.

A further experimental approach to deriving tissue structure information has been, whilst working with a limited acoustic frequency bandwidth, to record patterns of variation of echo intensity that occur with changes to the relative orientation between the tissue structure and the ultrasonic beam (6,7). Implementation of such an approach <u>in vivo</u> requires a purpose built scanning arm which physically constrains the transducer to point at the same volume of tissue whilst performing an isocentric B-scan (8). The technique has been detailed elsewhere (6) and will not be discussed further in this report. Initial clinical trials of this machine by Nicholas (9) indicated that the diffraction patterns associated with hepatic disease (see Figure 1) yielded a ninetysix per cent success for discriminating malignancy, including 11 cases of diffuse infiltration, from normal liver tissue. The technique also showed promise in characterizing

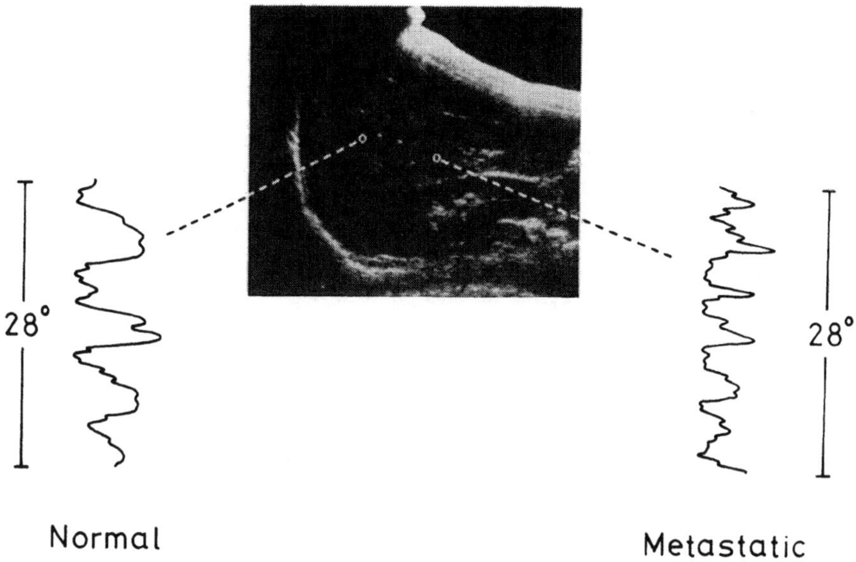

Figure 1. Liver B-scan and diffraction patterns from a patient with a focal metastatic deposit: right; focal lesion, left; normal liver tissue.

other non-malignant diffuse disorders such as cirrhosis and hepatitis. A subsequent study by Merton et al (10) on 52 patients with suspected thyroid disorders provided an eightyeight per cent success in discriminating between follicular adenomas, degenerating follicular adenomas and colloid masses.

It has been stressed (9) that the clinical value of this technique lies in its empirical ability to characterize various disease conditions, although it is still incapable of providing an absolute structural assessment of tissues except where a degree of anisotrophy exists, as in skeletal muscle. Complete structural evaluation can only be achieved by performing a three-dimensional investigation of the tissue - a situation which has been attempted with some success in vitro (11), but is still far removed from clinical implementation.

Although all patients were scanned under conditions of full inspiration, local tissue movement associated with the cardiac cycle could not be eliminated. It was, therefore, necessary to repeat every diffraction pattern associated with a specific tissue volume in order to check that the basic pattern remained unchanged. If sufficient correlation between successive scans was not achieved then those scans were eliminated from our study. In total, some 22 patients have proved to be impossible to scan by this technique simply due to tissue movements, and they have been eliminated from these studies. This represents a serious problem in diffraction scanning which will only be overcome by developing a system that can operate in 'real time', in the sense of collecting a diffraction pattern within a few milliseconds.

"Real time" applications

We have recently acquired a phased-array "real time" scanner which has been modified in a manner which allows the individual receive elements to be monitored (12). This prototype machine operates at 2MHz with arrays of 2 and 4cm length and utilises an 8-bit frame store for data acquisition. The arrays are comprised of 42 separate elements, half of which act as transmitters and half as receivers. Incorporated into the electronics are separate receiver circuits which time-select the received signals to those originating from a pre-determined region of the B-scan picture. An adjustable pair of crossed lines, superimposed on the real-time image, enables the operator to select the tissue region for investigation. The interference information (the energy received by each individual receiver) or angular diffraction pattern can thus be recorded (see Figure 2).

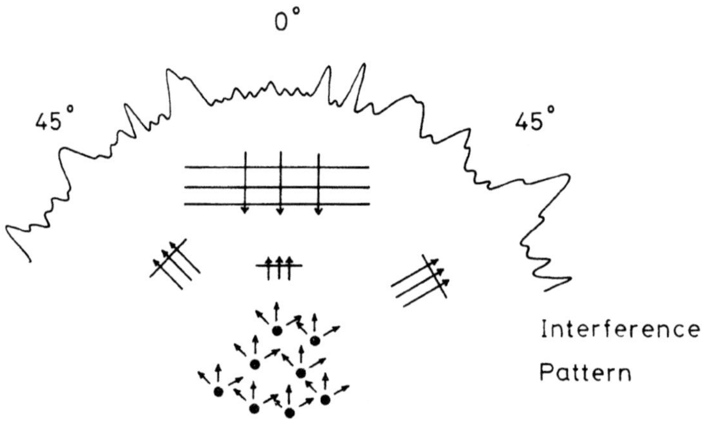

Figure 2. Schematic representation of angular scattering generating an angular diffraction pattern.

The volume of tissue investigated by this technique is determined by the transmit beam-width and pulse length at the chosen position within the sound field. For our system this limits the tissue volume investigated to approximately $20mm^3$. All the measurements (using the 2cm array) have been conducted on regions of tissue at distances between 3 and 12cm from the transducer face, thereby restricting the angular diffraction patterns to angles between 36° and 9° respectively.

To date, angular diffraction patterns have been obtained from both focal and diffuse diseases of the liver. In figure 3 the sector B-scans and associated diffraction patterns for various hepatic pathologies are portrayed. Although the patterns are obviously different in these four cases, they do not reflect the variations in pattern associated with a particular disease or patient. Retrospective analysis of the patterns by computer enables a rigorous quantitation of the data and subsequent correlation with disease state. In a preliminary study involving 19 normal volunteers and 46 patients with biopsy proven diffuse liver disease, the technique provided a 77% success rate in differentiating between normal and abnormal, and an overall success rate of 63% when classifying five types of diffuse hepatic disorder, namely: cirrhosis, chronic active hepatitis, fatty infiltration, diabetes and patients exhibiting both micro nodular cirrhosis and hepatitis. This later result is particularly significant when considering the inability of diagnosticians to achieve a similar success from simple visual assessment of the B-scan images.

Tissue Motion Studies

Originally, the fluctuation in the speckle representing tissue parenchyma, as visualized from "real-time" machines, was considered as an artefact of the system and consequently ignored. Latterly, an interest in these speckle changes has been fostered on the basis that such changes may be dictated by bulk acoustic properties of the tissue such as its rigidity or compressibility (7). In attempts to quantitatively monitor intensity changes in the spatial location of echoes, two techniques have been adopted to date.

The present author (13) has conducted a simple pilot study where the diffraction technique (described in the previous section) has been utilised to monitor changes in backscattered intensity as a function of time. As the diffraction technique has been described elsewhere (6, 9) it is sufficient to note that by time gating the A-scan pertaining to a selected direction a small tissue volume (typically 30 mm^3) can be investigated in reference to its backscattering characteristics. Any changes in tissue volume (with different scatterers entering and/or leaving the time-gate) will result in changes in the interference effects recorded. Although such measurements are unable to assess the degree of movement they are extremely sensitive to small scale local motions. Figure 4 illustrates two diffraction/movement patterns for differing liver conditions. The upper trace is associated with movements from within normal liver parenchyma whilst the lower trace is from a grossly enlarged cirrhotic liver. It is noticeable that changes in the diffraction information can occur over very short time scales compared with the heart cycle, and may be tentatively related to the general "pliability" of the tissues concerned.

An alternative method, adopted by Dickinson and Hill (14) is to monitor consecutive A-scans originating from a predetermined line scan as a function of time. It is then possible to correlate identical time-regions of the A-scans to yield a quantitative description of tissue motions. Both these approaches are in their infancy yet they indicate that this hitherto neglected aspect of ultrasonic diagnosis may well prove to be of paramount importance in differentiating between various pathological conditions.

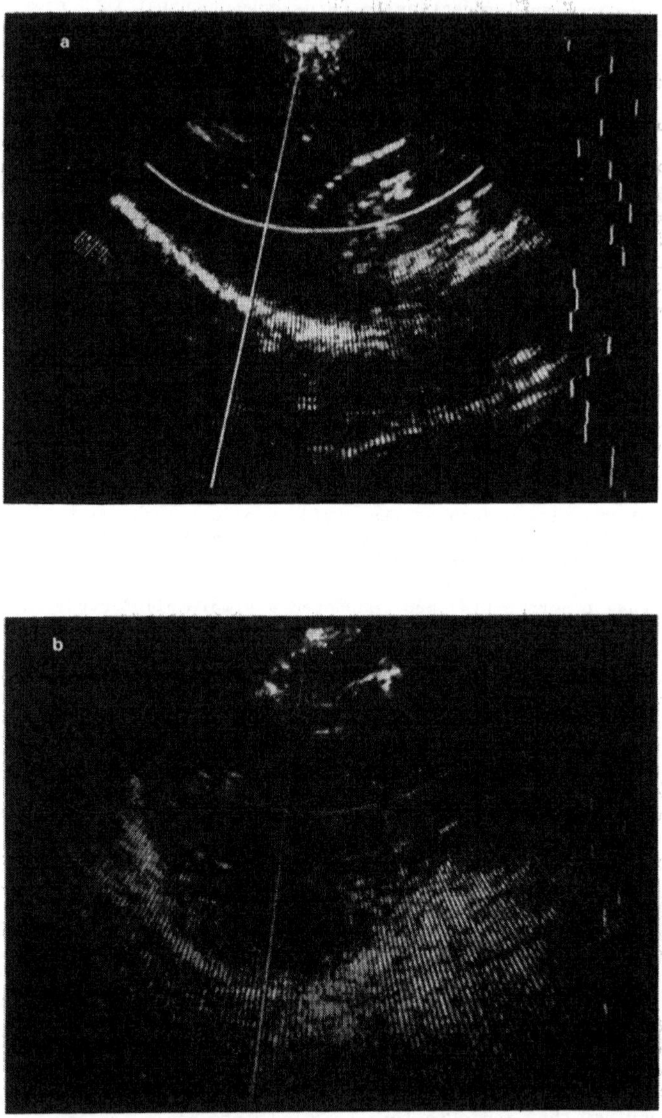

Figure 3. Sector B-scans and angular diffraction patterns for various hepatic pathologies :
(a) normal liver parenchyma
(b) a metastatic deposit within liver parenchyma

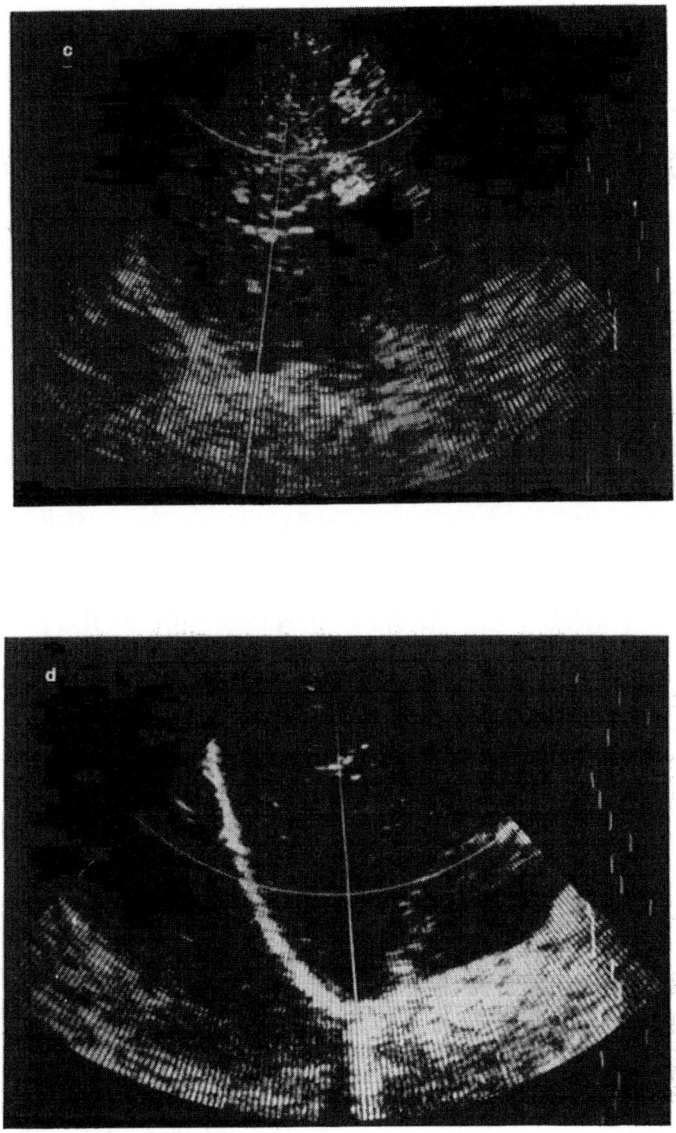

Figure 3. (c) diffuse alcoholic cirrhosis
(d) diffuse fatty infiltration

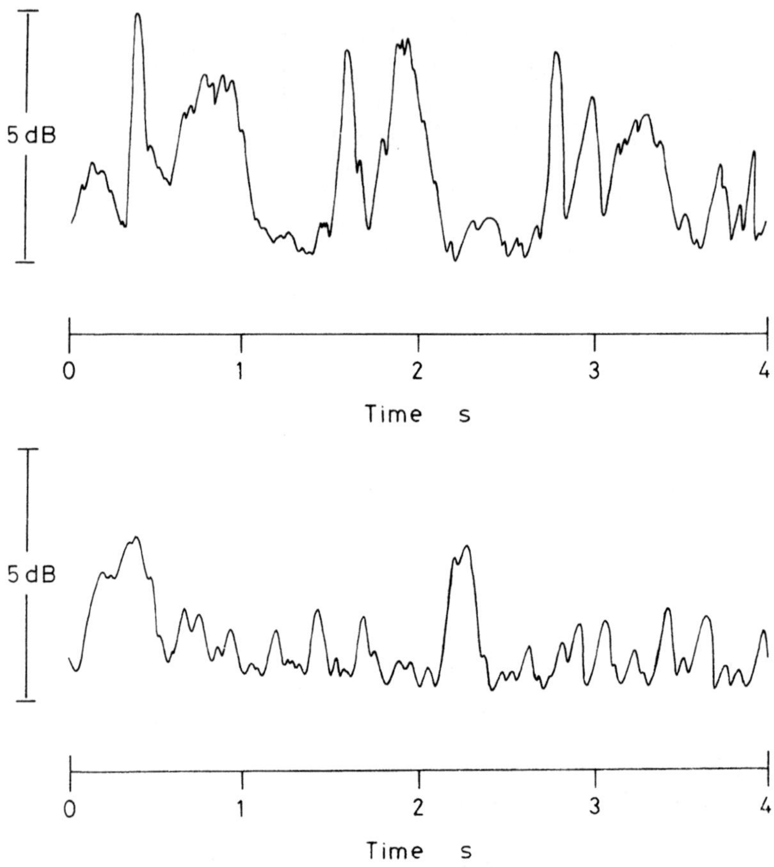

Figure 4. In vivo diffraction pattern changes as a function of time.
(a) Movements associated with normal liver parenchyma
(b) Movements from within a cirrhotic liver

Image Analysis and Enhancement

The information held in a full B-scan extends the A-scan into another dimension implying that techniques applied to simple A-scans could have more powerful applications when extended to two-dimensions. Hence there has been a recent growth of interest in applying image pattern and texture analysis to B-scans. The obvious drawbacks are the need for large data depositories, fast data capture and sophisticated computational facilities.

In our system (15) existing video storage devices, analogue scan conversion memories, have been used as temporary storage for subsequent transfer (at slow data rates) to permanent computer storage. Although this technique is relatively simple and inexpensive to implement it is limited by loss of all phase information originally present in the r.f. signal. Furthermore, this approach is also effected by the pre-processing of the echo information prior to scan conversion (16)

Once the B-scan data is stored in a computer two basic approaches can be implemented. One approach is aimed at manipulating the original image to form an "enhanced" version which can be better visually interpreted by the diagnostician, whilst the other implements two-dimensional feature extraction algorithms in a similar manner to those reported elsewhere in this volume.

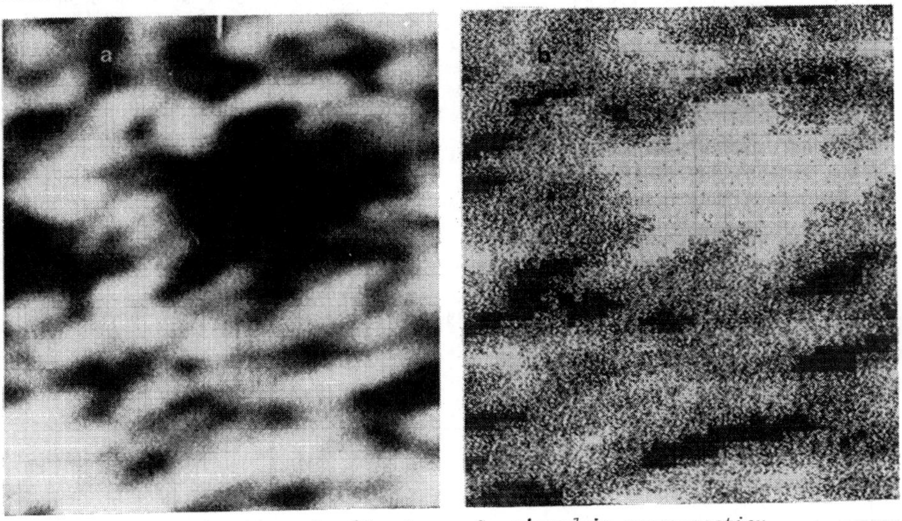

gure 5. (a) *Enlarged B-scan of a blood vessel, viewed in cross-section, displayed via a conventional analogue scan convertor*
(b) *Post-processed digital version of (a). Nicholas (16)*

Since analogue scan converters can provide a dynamic range of up to 40 dB, 100 grey-levels, it is not feasible to display the whole range of such information on our existing monitors. In order to demonstrate the usefulness of this extra information we have devised a system whereby small regions of the stored image (corresponding to a 3 cm square of tissue) can be selected, digitised to 8 bits and fed into a computer for subsequent image analysis and processing. By selective windowing and redisplay of the image details can now be perceived which were not apparent on the original 'full' image. Although windowing provides a useful technique it depends upon the operator to subjectively select the window which provides the information desired. This can lead to complications in that by viewing ranges of the information separately one can obtain a non-representative overall impression of the information content. One method for display of a wide range of information where the concentration of information is non-linear is to utilise full frame histogram equalisation. This forces the displayed grey-levels to occupy equal amounts of data, ie each grey-level is represented in the displayed image an equal number of times. Figure 5 compares the original data, displayed conventionally from the analogue scan converter, with a computer print out of the histogram equalised image with eight levels of grey. The display of the samples (pixels) in the digital image has been overcome by using a random dot display as a form of 'blurring' to remove edges. It can be seen that the blood vessel viewed in cross-section is better visualised using this technique especially as the original image did not display sufficient information at the vessel boundary to allow for immediate recognition.

Image analysis procedures implement two-dimensional feature extraction algorithms in a similar manner to those reported for the A-scan techniques elsewhere in this volume. We have utilised features extracted from second order statistics, ie the Grey level dependence matrix, Fourier spectral parameters and simple echo counting statistics. A discriminant analysis involving combinations of these features is thus performed on the appropriate data sets. A trial of 50 normal volunteers and 38 patients with biopsy proven diffuse liver disease have been analysed. Table 1 lists the discriminating ability of our technique for the five liver conditions examined. The overall 85% success in classification is extremely encouraging and justifies the use of the video signal for tissue differentiation.

These studies have indicated that quantitation of the images can greatly aid our present subjective evaluation, however, high technology computing facilitie have been implemented to provide the data extractions and analysis. The goal of such work will be the eventual selection of a limited set of features and appropriate analysis which can be performed by a microprocessor. It is then feasible that such techniques could be integrated into conventional ultrasonic imaging machines for on-line evaluation.

Table 1. Computer classification of diffuse hepatic disorders

BIOPSY \ COMPUTER	NORMAL	CIRRHOSIS	HEPATITIS	FATTY	CIRRHOSIS WITH HEPATITIS
NORMAL	47	1			2
CIRRHOSIS	5	16			1
HEPATITIS	1		4		1
FATTY				2	
CIRRHOSIS WITH HEPATITIS	1	1			6

Overall 85% success

CONCLUSION

Although we have approached the clinical implementation of ultrasonic tissue characterization from different aspects we cannot claim that any one technique is optimal. They have all proven to be of use in providing extra information to aid in diagnosing pathological disorders yet they must only be regarded as complementary to existing imaging techniques and not supplementary. The eventual success of such techniques can only be judged by their acceptance by diagnosticians generally. A goal that will only be achieved when the worth of such procedures is proven and manufacturers have been encouraged to incorporate them into marketable products.

REFERENCES

1. LIZZI F L, LAVIOLA M A and COLEMAN D J: Tissue signature characterization utilising frequency domain characterization. Proceedings of the IEEE Ultrasonics Symposium. New York, IEEE, 1976, pp 714-719.

2. SOMMER F G, JOYNT L F, CARROLL B A and MACORSKI A: Ultrasonic characterization of abdominal tissues via digital analysis of backscattered waveforms. Radiology (in Press).

3. THIJSSEN J M (ed) : Ultrasonic Tissue Characterization. Brussels, Stafleu, 1980.

4. Various Accompanying papers in this volume.

5. NICHOLAS D: Time-frequency-domain analysis: one dimensional phantom studies. Phys. Med. Biol. (in Press).

6. NICHOLAS D and HILL C R: Acoustic Bragg diffraction from human tissues. Nature 257: 305-306, 1975.

7. NICHOLAS D: Orientation and frequency dependence of backscattered energy and its clinical application. In White D N (Ed) : Recent Advances in Ultrasound in Biomedicine Vol.1, Oregon, Research Studies Press, 1977, pp 29-54.

8. HUGGINS R W and PHELPS J: Bragg diffraction scanner for ultrasonic tissue characterization in vivo. Ultrasound in Med. & Biol. 2:271-277, 1976.

9. NICHOLAS D: Ultrasonic diffraction analysis in the investigation of liver disease. Brit.J.Radiol. 52:949-961, 1979.

10. MERTON J, NICHOLAS D, HILL C R, GROVER S, QUEENAN M and COSGROVE D: Ultrasonic diffraction scanning of the thyroid. Ultrasound in Med. & Biol. (in Press).

11. NICHOLAS A W and NICHOLAS D: The analysis of three-dimensional tissue structures using ultrasound diffraction patterns. Phys.Med.Biol. 25:759, 1980.

12. NICHOLAS D, COSGROVE D O and PUSSELL, S: Real-time diffraction scanning. Brit.J.Radiol. 53:1025, 1980.

13. NICHOLAS D: The analysis of soft tissue movements. In Thijssen J M (Ed): Ultrasonic Tissue Characterization. Brussels, Stafleu, 1981, pp 281-285.

14. DICKINSON R J and HILL C R: Measurement of soft tissue motion using correlation between A-scans. Ultrasound in Med. & Biol. (in Press).

15. NICHOLAS D, BARRETT A, CHU J M G, COSGROVE D O, GARBUTT P, GREEN J, PUSSELL S and HILL C R: Computer analysis of grey-scale tomograms. In METHERELL A F (Ed): Acoustical Imaging Vol.8, New York, Plenum Press, 1980, pp 731-744.

16. NICHOLAS D: Pre- and post-processing of images held on temporary storage devices. In HILL C R and KRATOCHWIL A (Eds) : Medical Ultrasonic Images: Formation, Display, Recording and Perception, Amsterdam, Excerpta Medica, 1981, pp 61-68.

Discussion

C.R. HILL, D. NICHOLAS and D.O. COSGROVE

STATUS OF TISSUE CHARACTERIZATION WORK AT THE INSTITUTE OF CANCER RESEARCH AND ROYAL MARSDEN HOSPITAL

Question by: L. Masotti

What about the dependence of the diffraction pattern on the beam shape and frequency?

Reply by: D. Nicholas

These are features which play an important role in the production of our diffraction patterns. At present we have limited our *in vivo* studies to a specific frequency and transducer. Hence all the results are only relative. We are studying these equipment effects in vitro and by using computer modelling, but our current understanding is still in the preliminary stage and we hope to be more definitive in the near future.

Question by: M. Clement

You have not shown diffraction results with isocentric B scans of the breast. Is that because you had difficulties or because of lack of opportunity to carry out these clinical observations?

Reply by: D. Nicholas

We are still in the stage of developing a clinical breast scanning system. When we have sufficient clinical expertese in this field we will certainly apply this technique, although I envisage certain practical problems such as refraction of the ultrasonic beam at the breast surface.

Question by: D. Decker

I assume, that the texture analysis - whether it is done visually or by computerized, statistically methods - is strongly dependent on the US-equipment and the doctor, the setting of the attenuation and the TGC and last not least on the transducer.

Reply by: D. Nicholas

This can indeed be the case. However, it is necessary that one chooses texture parameters which are insensitive (relatively!) to such instrument settings. For example, 2nd order statistics are fairly independent (to a first order) to gain and/or TGC settings. Computer studies conducted at our institute suggest that the variations in equipment settings due to clinical preference do not adversely effect our results. Equipment features such as the transducer, however, will certainly change the results and we need to find a method for normalising such features.

Question by: J.M. Thijssen

How do you handle the lateral vs axial resolution and change of lateral resolution with depth, or in other words, is it not difficult to differentiate textures with so many parameters involved? Is the term even unambiguous?

Reply by: D.O. Cosgrove

When assessing texture by eye, one makes allowance for the change in lateral resolution with depth based on experience with particular, familiar transducers. Clearly this is subjective and only approximate. It is one of the variables which a formal more precise texture analysis must take account of, other being the transfer function of the electronic chain and of the intervening tissues, this latter being especially difficult to allow for.

Further discussions and remarks

The question as to which size of a tumour should be considered to be critical in respect to the danger of metastasis, was discussed.
A size of about 1 cm should be considered the upper limit. But often tumours of this size still are asymptomatic and diagnosis is not made. On the other hand in the liver a high risk group is formed by patients who have been treated for a primaire tumour and who are regularly screened for metastasis. In those cases the early detection of multifocal small tumours is very important and tissue differentiation may become a secundary goal of echographic diagnosis.

SPECTRAL ANALYSIS FOR TISSUE CHARACTERIZATION

J. Keuwez and S. Levi
Université Libre de Bruxelles, Hôpital Universitaire Brugmann, Dept. de Gynaecologie-Obstetrie, Service d'Ultrasonographie, Fondation Freserh, B-1020 Bruxelles, Belgium

Introduction

Spectral analysis is now applied in many fields thanks to the improved availability of digital systems. It is applied in the medical field, in particular for ultrasonic tissue characterization.
Some initial clinical studies were made by Lizzi et al. on the eye; later other organs were investigated (1)
In consideration of sampling theories (2) and the frequency range to be investigated, high sampling rates are required and, therefore, sophisticated A/D converters are needed.
Problems arise when one tries to apply spectral analysis to <u>in-vivo</u> tissue characterization of non-superficial organs. The causes of some of the problems encountered are determined and together with some aspects of the experimentation, will be described.

Matherial and Method

The equipment, which is more fully described elsewhere (3), was composed of a conventional ultrasonic scanner (Kretz Combison II, 2.5 MHz), a computer (Digital PDP 11/34), and a transient digitizer (Tektronix R7912). Echoes reflected from different parts of uterine and pelvic tumour tissues were recorded, and the data processed by Fourier analysis.
Only the amplitude spectra are considered in these experiments. For reasons, to be explained later, the date were analysed by comparing spectra corresponding to various tissues regions.
Although samples were obtained for random directions within the mass, only comparisons were made between spectra obtained along the same propagation path.

General

For a particular patient, the interactions between soundwaves and successive layers of intervening tissues (between the transducer and the target to be characterized) are far to complex to be known with any precision. Moreover, these interactions could vary significantly from patient to patient, and from one probe location to another one.
In view of this, our approach based on the comparison of the spectral content of internal echoes seems appropriate for characterizing ultrasonically non-superficial organs.
It is expected that the coefficient of attenuation in biological tissues (which is supposedly linearly dependent upon the frequency) is a signature of the tissue-state. Theoretically, the coefficient could be derived from the normalized logarithmic difference between a pair of spectra, pertaining to echoes situated on the same propagation path.
Nevertheless, the examination of only one of these amplitude spectra shows the difficulty in deducing the tissue attenuation when considering only a pair of echoes, however they were selected. The tissues to be characterized not only contain reflectors of irregular shape, but the reflected signals are also distorted by phase-cancellation effects and produce scalloped amplitude spectra.
The spectrum on one echo can not be considered as a deterministic characteristic of the tissue. A statistical approach should be adopted in order to extract useful information out of the echoes; as applied by Kuc (4) to differentiate liver diseases.
We tried to apply a similar method in our experimental studies. The equipment described above did not allow us to record a sufficiently large amount of data for a good statistical evaluation; however these experiments enable us to elucidate some aspects:

1. The time-delay between analyzed samples from the high-frequency A-scan signal should be long enough to experience sufficient attenuation effects corresponding to depth penetration. The necessary delay depends upon the frequency band and tissue. However, if it is too long, the echoes may originate from places where the transducer characteristics are greatly different.
   It should be wise to take into account the field characteristics, but they have not been well-defined. Indeed, even if known in a non-attenu-

ating medium, extrapolation to an attenuating, inhomogeneous medium may not prove valid.
A well characterized transducer should partially solve this problem; however, besides that, it is not always possible to place the target tissues in the optimum field region.

2. Due to inhomogeneities and discontinuities in the medium, the frequency signal is composed of a multitude of small and large echoes. When experiments are conducted in-vivo, the total duration of data acquisition must be minimised to avoid tissue motion effects. Consequently it is impossible to repeatly adjust the time-window position and the amplifier gain of the transient digitizer to get the optimum signal for digital conversion.
Furthermore the segmentation is not ideally done due to the overlapping of non-equal amplitude echoes. It results in a waste of data and less acceptable information available for analysis.
From these considerations, we found the segmentation method, proposed by Kuc (4), difficult to realize in the experiments described here, where we are attempting to derive a good estimate of the attenuation coefficient.

3. The appropriate averaging of the normalized slopes of the straight lines which best fit the difference of amplitude spectra gives an estimate of the coefficient of attenuation. It should be emphasized that positive as well as negative values could be obtained. This straight line fit constrains the spectrum between limits which are chosen in such way that outside these limits, the spectral content may be considered as negligible. Furthermore, these limits are arbitrarily chosen and changing them can drastically change the slope of the line (lower or greater slope). Although it is to be hoped that the mean slope is not influenced.
The problem of the spectral limits is related to the fact that the spectra are scalloped. Spectra with significant distortion should be neglected. This new limitation again affects the mean estimation of the attenuation coefficient.
A preliminary averaging of spectra corresponding to echoes at the same depth, on closely related propagation paths, could produce smoothed spectra. However, as mentioned above, intervening tissues can show different properties from one point to another. Averaging may thus lead to

erroneous results.

The attenuation coefficient obtained in such a way is too sensitive to experimental parameters. We propose that there should exist functions g such that :

$$G(d,y) = \int_0^y g(A(d,x))dx$$

where $A(d,f)$ is the spectral amplitude at a frequency f, and d the depth penetration : these characteristics are related to the attenuation coefficient and less dependent on external factors (5).

CONCLUSIONS

Preliminary experiments were conducted with a view to characterizing non-superficial tissues using spectral analysis. These constitute a generalization of a method we developed earlier (6).

The processing of our initial data shows that a statistical approach is more appropriate for achieving the characterization. The purpose is not to determine the attenuation coefficient absolutely, but to characterize tissues through the use of any parameter which is a function of the attenuation. In other words, we are proposing a method where spectral contents of discrete echoes are compared. However, it is not absolutely necessary to extract the time attenuation from such a comparison; other related parameters could be more useful.

Due to difficulties of sampling, other experimental conditions are being developed.

REFERENCES

1) Lizzi, F., Katz, L., St. Louis, L. and Coleman, D.J. (1976): Applications of spectral analysis in medical ultrasonography. Ultrasonics 14, 77-82.
2) Papoulis, A. (1977): "Signal Analysis". New York, McGraw-Hill.
3) Levi, S., Keuwez, J. and Bredael, Y. (1978): Tissue characterization in vivo: a semi-automatic method. In: "Recent Advances in Ultrasound". Asim Kurjak, Ed. Amsterdam, Excerpta Medica, 36-47.
4) Kuc, R. (1980): Clinical application of an ultrasound attenuation coefficient estimation technique for liver pathology characterization. IEEE Transactions on Biomedical Engineering, Vol. BME-27, Nr. 6,
5) Levi, S. and Keuwez, J. (1981): Tissue characterization of the kidney. In: "Diagnostic Ultrasound in Urology and Nephrology" (Ed. H. Watanabe, J. Holmes, H. Holm, B. Goldberg) Igaku-Shoin, 61-65.
6) Levi, S. and Keuwez, J. (1979): Tissue characterization in vivo by differential attenuation measurements. In: "Ultrasonic Tissue Characterization II". Ed. M. Linzer (NBS Special Publication) 525: 121-124.

Discussion

J. KEUWEZ and S. LEVI

Tissue characterization: statistical approach

Question by: L.F. v.d. Wal

Are you aware, that without any processing your measurements are deformed by transducer characteristics, beam-forming, propagation effects, etc. If so, what is the value of analyzing single measurements in frequency domain?

Reply by: J. Keuwez

I am aware of that problem but the beam profile and its spectral contents along the propagation axis are influenced by the structure and the properties of the intervening tissues which are unknown and which could be different from patient to patient. At this stage I do not have a good solution for that problem.

Question by: I. Bredael

How do you determine the parameters for the signal window construction in function of the signal frequency content?

Reply by: J. Keuwez

I followed the procedure described by Kuc, and my experiments did not lead to another choice.

Question by: L. Masotti

Would it be possible to increase the statistical meaning of the spectral amplitude calculation by varying randomly the position of the time window along an appropriate time interval included in the equivalent time duration of the mass to be studied.

Reply by: J. Keuwez

If you proceed in that way, you are not sure to take independent samples which is not compatible with the statistical analysis which has to be performed.

Question by: H.G. Trier
How did you register the foetal rapid eye movements (REM)?

Reply by: J. Keuwez
We use a scanner (linear array) which gives high-resolution images to visualize the eye movements. The scanning is made in a frontal plane at the eye level.

Question by: K-M. Irion
You want to use the energy distribution to estimate the slope of attenuation instead of the "center of gravity" method. I think, that the "center of gravity" method is only a derivation of the energy distribution as a function of time and frequency, isn't it?

Reply by: J. Keuwez
Yes, but I said that the calculation of the slope of attenuation is dependent of various factors such as the bandwidth, and I think it should be better to use a method which is not dependent of these factors. A way to achieve that could be through the energy distribution, and of course the "center of gravity" is one of the parameters we can derive from it.

Question by: K-M. Irion
Which algorithm do you use for the segmentation of the RF signal?

Reply by: J. Keuwez
I was using the algorithm proposed by Kuc.

Further discussions and remarks

The difficulties connected to working in in vivo situations were mentioned. One of these is the interference from intervening healthy tissues, which is different for every patient. Therefore, the beam characteristics of the sound energy arriving at the examined tissue volume may be different for every patient and for every direction chosen in a particular patient. The solution may be partly found in the averaging and in taking relative measurements at various depths along a single propagation direction.

ULTRASONIC CHARACTERIZATION OF ATHEROSCLEROSIS: IN VITRO FINDINGS AND PERSPECTIVES FOR IN VIVO APPLICATIONS

L. Landini, E. Picano, A. Distante, E. Falchi and A. Benassi
Centro per l'Automatica 'E.Piaggio', Faculty of Engineering, Pisa
C.N.R. Institute of Clinical Physiology at the University of Pisa and Institute of 1st Medical Pathology of the University of Pisa, Italy

ABSTRACT

In attempts to non-invasivly characterize normal and atheroslerotic arterial walls, we undertook a study in which two ultrasonic transmission indices were derived to differentiate pathological samples of human aorta.
These indices were explored both as a function of frequency and as integrated values over the frequency range 1-10 MHz.
It has been observed that, using both the employed transmission indices, it is possible to distinguish atherosclerotic from normal aortic walls, and to cluster atherosclerotic processes according to different degrees of severity.
The structural basis of the recorded differences is discussed in the present paper, where gross pathology, histology and biochemistry of the aortic samples have been correlated with ultrasonic data.

Introduction

The need for development of ultrasound systems which can provide useful information about peripheral arterial disease is unquestionable. A working group of the "National Heart Lung and Blood Institute" in September 1978, put as its first priority, in its final recommendations, "the research (....) of non-invasive systems for the detection, quantification and characterization of atherosclerotic lesions" (1).
The present study was designed to determine "in vitro" whether specific alterations in acoustic parameters, potentially useful for diagnosis, characterize the atherosclerotic processes by providing a quantitative assessment of severity. The approach used was based on two ultrasonic in-

dices which exploit transmitted ultrasound. The measured indices yield information on both the impedance and geometry (intimal surface irregularities) of the specimens under study (2). The experimental perspectives in view of in vivo applications are also discussed.

Methods and Instrumentation

The work was carried out on 100 samples of normal and atherosclerotic aortic walls.

A. Pathological characterization.

The arterial wall specimens were taken out during autopsy and fixed in 5% formalin solution, for a constant time interval (2 days). Each aortic wall specimen (macroscopically categorized as normal, fibrofatty, or calcific) after ultrasonic study was cut into two symmetrical parts: one underwent a histological analysis, the other a biochemical study (hydroxyproline and calcium dosage).

For histology, a standard ematoxilin-eosin and a Weigert-Van Giesen coloration was made for each sample. The classification was made according to generally accepted criteria (3). Macroscopic and microscopic classification - made independently - always resulted in full agreement.

B. Ultrasonic characterization.

Two different indices, related to ultrasonic attenuation, were measured : "transmission index" and "Integrated Transmission Index".

1. "Transmission index" (K): Fig. 1 shows the system employed to obtain "K". It consists of a gated burst generator, (Wavetek model nr. 616) externally triggered, a dual trace oscilloscope, two transducers (transmitter and receiver; Aerotech, model alpha) with 10 MHz center frequency and 1 cm diameter, and a tank for transducer micrometer displacement. Data collection was made manually. The transmitted signal consisted of sinusoidal wave-bursts (2 μsec. in time) in the frequency range of 1-10 MHz. Ten values of frequency were chosen to test the tissue. The value reported at each frequency represents the average of three measurements taken at the stated frequency and at 50 KHz above and below the stated frequency. The transmission index is expressed (in percent) in relation to the reference system.
2) "Integrated Transmission Index" (ITI): Measurements were made by a single (10 MHz, Aerotech model alpha) transducer acting as transmitter and re-

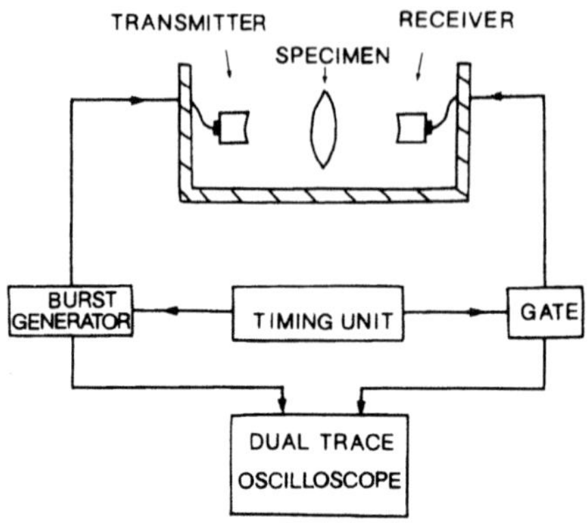

Figure 1: Block diagram of the ultrasonic instrumentation used in this work

ceiver of the echo produced by a specular reflector positioned behind the specimen under study. In this approach the frequency dependence of the echo amplitude that has passed twice through the specimen under study is measured. In these measurements the electrical source consisted of a broadband pulser-receiver (Panametric model 5052PR). Data collection was performed automatically. Under the control of a timing unit, broadband ultrasonic pulses are digitized by a transient recorder (prototype implemented in our Institute) at a rate of 10 µsec., with 6 bits of amplitude resolution, and stored in a buffer of 1024 words of 16 bits each. Data from the buffer are continuously transferred to a minicomputer (HP - 21 MX) for signal analysis. The analysis has been performed by using the FFT algorithm to obtain the frequency spectrum in the range 1-10 MHz. In this range the integrated transmission index has been calculated. The tissue specimen was placed on a sample holder, 5 cm from the transmitting transducer with the intimal surface faced towards the transmitter. The incident beam was kept normal to the aortic wall sample. The transmission measurements have been performed using distilled water as a coupling medium.

C, Biochemical characterization.

One half of the aortic sample was also characterized for both hydroxyproline (a "marker" aminoacid for collagen) and calcium content. Hydroxyproline was assayed, after proper hydrolysis, with a Beckman aminoanalyzer, and its values expressed as micromoli per gram of dry weight. The highest findings occured in the fibrofatty subset (10.4 ± 2, mean ± SE) as compared to normal (7.9 ± 1.8) and calcific (2.2 ± 0.5) subsets. Calcium was assayed with an atomic absorption based instrument, and its values were expressed as nanomoli/gram of dry weight. It produced highest results in the calcific subset (22.6 ± 3) as compared to normal (1.5 ± 0.2) and fibrofatty (9.1 ± 1.4) subsets.

D. Statistical analysis

The mean value and the standard error were evaluated for each normal and pathological subset.

In vitro results

Ultrasonic results are based on the analysis of both the slopes and absolute values of K and the absolute value of ITI.

As far as "K" is concerned, the three pathological subsets (defined above) differ for both absolute value and slope. The K value is highest in the normal (NL) group, lowest in the calcific (CL) group and intermediate in the fibrofatty (FF) group (Fig. 2). At higher frequencies the rate of slope decrease is minimal for normal walls, and increases for the fibrofatty group. In calcific walls the slope is constant over the frequency range considered.

Also with "ITI" the distinction between the subsets is very clear cut (Fig. 3).

Discussion of the results

The results of our study indicate that - with both "K" and "ITI" it is possible to distinguish atherosclerotic lesions from normal aortic walls and to discriminate different degrees of severity in the atherosclerotic process (fibrofatty from calcific plaques). This alteration in acoustic behaviour by atherosclerotic tissue may reflect intrinsic changes in biochemical and/or mechanical (geometrical) properties of the aortic wall.

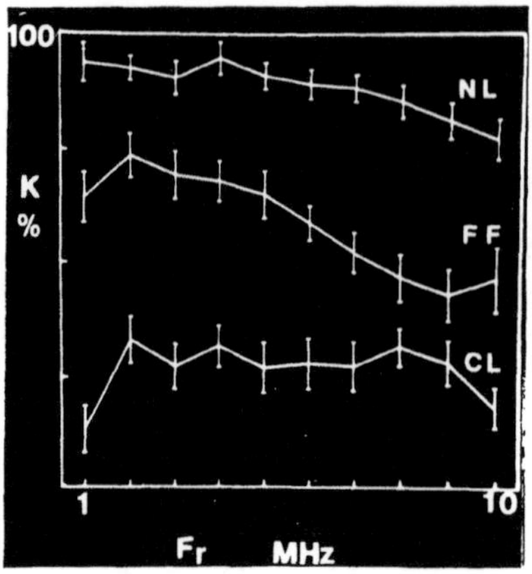

Figure 2: Transmission/frequency plots for three subsets of aortic walls: NL = normal walls; FF = fibrofatty walls; CL = calcific walls

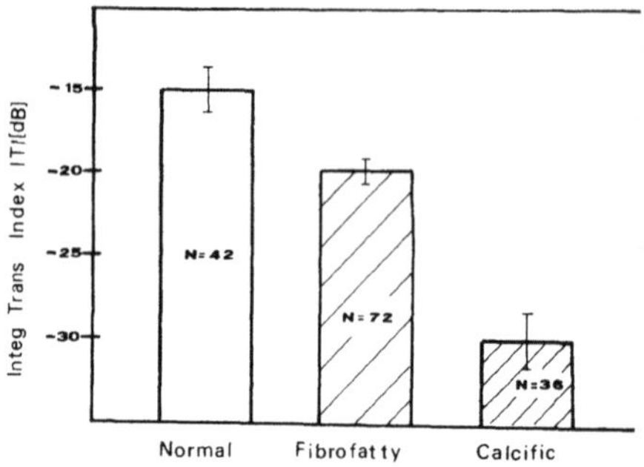

Figure 3: Integrated Transmission Index for three subsets of aortic walls; N is the number of samples in each subset

The decreased ultrasonic transmission in atherosclerotic tissue appears to be related to increased deposition - as documented by biochemical analysis - of highly absorbing materials, such as collagen (noticable in fibrofatty walls) and/or calcium (in calcific walls).

Field and Dunn (4) have outlined the theoretical basis for suspecting that ultrasonic impedance and attenuation (and consequently K and ITI, which are inversely related to attenuation) in soft tissue are determined primarily by structural components which contain collagen. More recently, it has been demonstrated that collagen appears to be the principal determinant of the elevated attenuation detected in regions of myocardial infarctions (5).

As far as calcium is concerned, it is known to have one of the highest impedances among biological molecules; therefore it is likely that its increased presence in calcific walls accounts for the very low K and ITI values relating to this particular subset. In pathological specimens another possible reason for decrease in transmission is a change in the mechanical properties (geometric surface irregularities: "roughness"). When compared with normal aortic wall - which is almost perfectly smooth - there is an increase in roughness in pathological aortic walls (roughness is known to increase reradiative energy loss, thereby decreasing transmission). It is difficult, however, to discriminate the relative importances of increased impedance and roughness in determining the K variations.

PERSPECTIVES FOR IN VIVO APPLICATIONS

Some basic experimental problems must be solved before any attempt to transfer the obtained results to an "in vivo" setting :

a. Extraction of information from reflected signals.

In future studies "in vitro" ultrasonic parameters exploiting reflected ultrasound (these are potentially most amenable to "in vivo" applications) should be used. Such studies should also make clear which of the possible parameters extractable from the reflected signals (integrated backscatter vs diffraction method vs Hilbert transform) is best suitable in our experimental model (atherosclerosis detection) (5,6).

b. Compensation procedures.

The importance of knowing the effects of the interposed tissues on the signals reflected from deep target structures comes from both theoretical considerations and experimental work (7). In fact, the spectrum of a signal coming back from a deep reflecting organ is determined partly by target scattering cross section and partly by attenuation coefficients of interposed tissue layers between the target organ and the transducer. Only a direct measurement of these attenuation coefficients would allow one to add to the received echo the amplitudes of the spectral components attenuated by interposed tissue ("compensation") in order to attempt quantitative studies on the target organ. The measurements of the attenuation coefficient will be carried out for different incident angles between the ultrasonic beam and the vessel and in subjects with different thickness of the cutis.

As an example, for measuring the extracranial carotid artery, a potential target-organ, compensation should be made for the tissues which lay - in cervical region - between cutis and carotis.

REFERENCES

1. U.S. Department of Health, Education and Welfare (1979): Assessment of non invasive instruments for visualizing plaques in peripheral arteries: summary of the Working Group. NIH, Publication no. 79-1963, pp. 1-7, August 1979.
2. Landini, L., Picano, E., Distante, A., Falchi, E., Pulella, G., Salvatori, M. and Benassi, A. (1981) : Studio della dipendenza dei coefficienti di transmissione e di backscattering dei processi di aterosclerosi in vitro. Pp. 337, Atti del VI Congresso SISUM.
3. Picano, A., Distante, A., Landini, L., Pulella, G. and L'Abbate, A. (1981) : Caratterizzazione ultrasonica in vitro mediante backscattering di pareti aortiche normali e aterosclerotiche. Pp. 331, Atti del VI Congresso SISUM.
4. Fields, S. and Dunn, F. (1973): Correlation of echographic visualizability of tissue with biological composition and physiological state. J. Acoust. Soc. Am. 54: 809-812.
5. Mimbs, J.M., O'Donnel, M., Bauwers, D., Miller, J.W. and Sobel, B.E. (1980) : The dependance of ultrasonic attenuation and backscatter on collagen content in dog and rabbit hearts. Circ. Res., Vol. 47, nr. 1, 49-58.
6. Nicholas, D. (1979): Tissue characterization in vivo by ultrasonic backscattering analysis. In: Ultrasound in tumour diagnosis. Eds. C.R. Hill, V.R. McCready and D.O. Cosgrove. Chigago, Pitman Medical Publishing Co. Ltd., pp. 258.
7. Miller, J.G.: Unpublished data.

Discussion

L. LANDINI, E. PICANO, A. DISTANTE, A. BENASSI

Ultrasonic characterization of atherosclerosis: in vitro findings and perspectives for in vivo applications

Question by: J.M. Thijssen

What is the kind of processing you applied to set such a broadband estimation of the transmission index?

What type of transducer did you use?

Reply by: L. Landini

We evaluated the efficiency $\mu(\nu)$ of the measurement system by the following relationship:

$$\mu(\nu) = \sqrt{\frac{v_r(\nu)}{v_t(\nu)}}$$

where: $v_{r(\nu)}$ and $v_{t(\nu)}$ are respectively the ν-components of the spectra of the received and transmitted signals, and we referred the measurements to this efficiency curve.

We used the following transducer type: Aerotech, model α, 10 MHz maximal frequency, 5 cm focus.

Further discussions and remarks

Some discussion was held on the miniature PVDF transducer used in intravascular studies as a receiving transducer.

ULTRASONIC TISSUE CHARACTERIZATION BY TEXTURE ANALYSIS

R. A. Lerski, P. Morley, M. J. Smith, E. Barnett, P. R. Mills, G. Watkinson
and R. N. M. MacSween
Western Infirmary, Glasgow, United Kingdom

1. INTRODUCTION

Early work by our group (3,4) was successful in showing that the application of methods of texture analysis to digitally recorded radio-frequency (RF) ultrasonic A-scan signals could provide parameters of diagnostic value in diffuse liver disease. The method has been refined and the requirements for its implementation studied in some detail. It is now believed that a clinically useful technique can be specified provided agreement can be reached on methods of standardisation.

2. METHOD

Although most modern ultrasonic B-scan apparatus store their images in digital format there is at present no evidence that such data may be used for ultrasonic tissue characterisation studies. Indeed recent work (6) has shown demodulated A-scan data to be inferior to RF data for texture analysis purposes. Although the factors of digitization accuracy, compression and two dimensional information are involved it is believed that such B-scan data is unlikely to be immediately useful in this context. The method developed in our laboratory has relied on the analysis of RF data since it is the most basic data available.

The RF ultrasonic signals from the transducer are passed through a wideband preamplifier to a transient recorder (Datalab DL920) where they are digitized at a rate of 20 MHZ/word to a precision of 8 bits and stored in a 4K buffer store. The data may then be transferred to the main memory of a Data General Nova 3 dedicated minicomputer for processing or for storage on floppy or cartridge disk. Triggering of the recording process is with reference to a digital marker adjustable within the region of interest of the B-scan. The size of the buffer store and digitization rate determine the 'length' of A-scan which may be recorded - for 4 K store and 20 MHZ rate the equivalent length is 154 mm. This system may be incorporated in an existing B-scan installation with minimal modification, the only signals necessary being the RF from the

---

Support for this work was provided by the Computer Research Committee, Scottish Home and Health Department.

transducer and a suitable digital trigger. For standardization purposes only the transducer and transmitter will affect the measurement.

Figure 1 illustrates the typical appearance of a digitally recorded RF A-scan signal from the liver. The structure can be seen and the sense in which it has a texture appreciated - there is a repetitive occurrence of local patterns of small and large peaks. It is reasonable to suppose that a change in the scattering structure of the liver - with a disease process, for example - might alter the textural appearance of the signal. Indeed there is evidence to support this view from the tissue differentiation observed in conventional grey scale B-scanning. Whether this is detectable, consistently measurable and of diagnostic value has been the subject of our investigations. Further questions of importance are the amount of data necessary to characterize a liver and the mode of data collection.

In work performed so far seven RF A-scans of length 77 mm were recorded from three sites or directions through the liver. The directions of the beam were selected to avoid major vessels or bile ducts and care was taken that the beginning and end of the signal lay within the liver. All data has been collected using a Fischer NE4200 grey-scale B-scanner with a Diagnostic Sonar 2.5 MHZ, 19 mm diameter, long internal focus transducer. No swept gain is applied to the recorded RF signal.

3. TEXTURE ANALYSIS

Three separate methods of texture analysis extracted from the literature have been applied to the RF signal data.

(a) Max-min texture analysis

This method of texture analysis adapted from Mitchell et al. (9) operates through the application of a digital smoothing algorithm to blocks of RF data (1K points). A count is made of the number of peaks that remain in the signal as an increased smoothing power is applied. Texture parameters (five in number) are defined as the ratios of peak counts after preset smoothing to that in the raw signal. An important property of this method is that with the use of logarithmic amplitude and smoothing threshold scales the texture measurements can be made invariant of gain changes in the recording system. A further advantage of the Max-Min method is its computational simplicity.

(b) Grey-tone spatial dependence texture analysis

This method is based on work described by Haralick et al. (2) and Pratt (10). A grey-tone spatial dependence matrix is constructed whose elements P(i,j) represent the number of occurrences of quantized values i and j separated by a particular correlation length L in the 1K block of RF data. All our data have been derived for a correlation length of 10 points equivalent to 0.4 mm in tissue. Although the data is digitized to 8 bit accuracy (256 levels) this is reduced to 5 bits (32 levels) to prevent unmanageable matrix sizes and to achieve reasonably large occupancy levels for the matrix elements. This latter condition is necessary to obtain statistical confidence in the matrix, but may lead to a loss of sensitivity to low amplitude textural changes.

Once the matrix is constructed parameters may be defined to describe its structure, e.g. spread about the diagonal, amount of local variation and so on. Full details have been given previously (5) and sixteen parameters defined.

(c) Grey-level difference texture analysis

Weszka et al. (11) described a method where a histogram H(k) was defined whose levels represent the number of occurrences of differences k = i - j in quantized signal values again separated by a correlation length L of 10 points. Four parameters (see (8)) may be derived from this histogram.

The three methods of texture analysis implemented so far yield a total of 25 parameters that describe the 1K block of RF data scattered from the liver. The usefulness of each parameter must be assessed for separating normal from pathologically abnormal liver tissue and for distinguishing between categories of disease.

4. PARAMETER ASSESSMENT

The usefulness of single parameters and combinations of parameters in separating disease states of the liver has been evaluated using multivariate analysis of variance (MANOVA) and discriminant analysis, the software being based on that given by Cooley and Lohnes (1).

Suppose two disease groups are present, A and B. Firstly in a procedure analogous to a Student t-test, univariate F-ratios are derived for each parameter to determine its value in separating the group means.

A high F-ratio supports the rejection of the null hypothesis, viz. that the group means are the same. The probability value (p value) that the group means are the same may be determined from tables of the F statistic.

Up to this point the same results could be obtained from a Student t-test. However, it is possible to extend the MANOVA method to multiple parameters and thus determine the extent to which the groups may be separated using combinations of measurements. This is done using a Fisher-type linear discriminant function formed to take account of parameter variance and covariance. A classification of the patient groups with such discriminant functions can indicate their separability using the parameters under study and the appropriate F-ratio the statistical significance of the separation in group means.

A good way of visualising multivariate discriminant analysis is to regard the measured parameters as features in a multidimensional feature space. An individual patient occupies a position in this space defined by his parameter values. It is hoped that in the case of a set of useful parameters patients of similar disease state will be grouped together in this space. The analysis starts by investigating group separations in single dimensions and proceeds to find the n-dimensional space where group separation (defined by an n-parameter discriminant function) is optimised.

5. TYPICAL RESULTS

Detailed results achieved so far have been presented previously (5-8). Using a combination of average echo amplitude and texture parameters in a series of 59 normal control livers and 45 histologically confirmed alcoholic cirrhotic and fatty livers the results presented in Table I were achieved.

Table I. Multiple parameter discriminations

Discrimination	F-ratio (p value)	Classification accuracy %
Normal/Fatty	$F^{3}_{111} = 58$ (0.001)	94
Normal/Cirrhotic	$F^{3}_{138} = 15$ (0.001)	83
Fatty/Cirrhotic	$F^{4}_{114} = 8$ (0.001)	84

It has been found that optimal classification accuracies are achieved using three or four parameters, this situation arising because of the often high degree of correlation between the various texture parameters. In a multiparameter discriminant function the addition of more parameters produces no improvement to the results or sometimes a deterioration. This behaviour has recently been investigated in detail using multiple covariance analysis (8).

In pattern recognition terms the results presented in Table I are from a training set of cases. The investigation has been extended to a further test set of 45 alcoholic liver cases, with the results presented in Table II simply for the normal/abnormal discrimination.

Table II. Training and Test evaluation

TRAINING	Normal/Abnormal	84% accurate
TEST	New group of 45 alcoholic liver cases	87% accurate

It can be seen that no loss of discriminant accuracy occurs in these cases.

In summary, the results up to now show RF texture analysis of ultrasonic signals to be capable of providing useful diagnostic data in diffuse liver disease. The pathologically abnormal liver can be distinguished from the normal which could lead to a reduction in the requirement for needle biopsy of the liver. Further it may be possible to discriminate between different disease states of the liver - an entirely new development with ultrasound.

6. DATA REQUIREMENTS

In terms of the implementation of the technique in clinical practice it would be a great advantage if data other than RF could be used. Present commercial equipment stores and presents ultrasonic data in a demodulated format. Digital scan convertors have available compressed and frequency limited data which might be utilized for tissue characterisation.

The availability of blocks of recorded RF data means that the question of the usefulness of demodulated data can be attacked in two ways. Firstly the relative diagnostic importance of certain frequency groups in the recorded signal may be investigated by Fourier spectra partitioning (7). Although this technique, in itself, does not compare favourably with the use of texture analysis it does demonstrate that a restriction to low frequency information leads to a loss of discriminant accuracy. That is, diagnostically useful data is present in the high frequency part of the scattered ultrasound spectrum.

Secondly it is possible to construct a simulated demodulated A-scan from the recorded RF by frequency filtering (6) and perform the texture analysis on this. Again it is found that a significant loss of discriminant accuracy takes place, manifested as an increase in false negatives. In the training set of 59 normal controls and 45 abnormal alcoholic livers discussed above, RF texture analysis only could achieve a 83% overall classification accuracy (False Positive FP = 16%, False Negative FN = 18%, $p = 0.001$) whereas demodulated texture analysis could only achieve 73% (FP = 22%, FN = 33%, $p = 0.001$).

7. DISCUSSION

Work so far has shown that RF texture analysis is a diagnostically useful technique in diffuse liver disease. A full clinical evaluation should now take place and is being planned in our centre.

The one existing problem to be settled is that of standardization, since our experience is that gel based scattering phantoms are not reliable enough for tissue characterisation standard purposes. What is required is a completely stable scattering structure in which one can have total confidence as a texture baseline.

REFERENCES

1. Cooley, W.W., and Lohnes, P.R. (1971). "Multivariate Data Analysis, Wiley," New York.

2. Haralick, R.M., Shanmugan, K., and Dinstein, I. (1973). Textural features for image classification. IEEE Trans. Systems, Man, Cybern. SMC-3, 610-621.

3. Lerski, R.A., Barnett, E., Morley, P., Mills, P.R., Watkinson, G., and MacSween, R.N.M. (1979). Computer analysis of ultrasonic signals in diffuse liver disease. Ultrasound in Medicine and Biology 5, 341-350.

4. Lerski, R.A., Morley, P., Barnett, E., Macsween, R.N.M., Mills, P.R., and Watkinson, G. (1980). Texture analysis of ultrasonic signals in the investigation of liver disease. In: "Ultrasonic Tissue Characterization - Clinical Acievement and Technological Potentials". Editor J.M. Thijssen, Stafleu (Alphen, Brussels), 257-265.

5. Lerski, R.A., Smith, M.J., Morley, P., Barnett, E., Mills, P.R., Watkinson, G., and MacSween, R.N.M. (1981). Discriminant analysis of ultrasonic texture data in diffuse alcoholic liver disease. Ultrasonic Imaging 3, 164-172.

6. Lerski, R.A., Morley, P., Mills, P.R. and Watkinson, G. (1981). Texture analysis of ultrasonic signals - a comparison of the use of radiofrequency and demodulated A-scan data. Ultrasonic Imaging (in press).

7. Lerski, R.A., Morley, P., Barnett, E., Mills, P.R., Watkinson, G. and MacSween, R.N.M. (1981). Ultrasonic characterisation of diffuse liver disease - The relative importance of frequency content in the A-scan signal. Ultrasound in Medicine and Biology (in press).

8. Lerski, R.A., Morley, P., Mills, P.R. and Watkinson, G. (1981). Texture analysis of ultrasonic signals - a comparison of methods through multiple covariance analysis. Submitted to Ultrasound in Medicine and Biology.

9. Mitchell, O.R., Myers, C.R. and Boyne, W. (1977). A max-min measure for image texture analysis. IEEE Trans. on Computers C26, 408-414.

10. Pratt, W.K. (1978). "Digital Image Processing", Wiley, New York, Chapter 17.

11. Weszka, J.S., Dyer, C.R. and Rosenfield, A. (1976). A comparative study of texture measures for terrain classification. IEEE Trans. Systems, Man. Cybern. SMC-6, 269-285.

Discussion

R.A. LERSKI, M.J. SMITH, P. MORLEY, E. BARNETT, P.R. MILLS, G. WATKINSON and R.N.M. MaCSWEEN

Ultrasonic tissue characterization by texture analysis

Question by: J.M. Thijssen
I would like to ask one of the clinicians a more general question. Do you think that it is possible to define a minimum volume necessary for tissue differentiation, or do you think we even need to scan as much as possible, maybe even the whole organ?

Reply by: P. Morley
It is not possible to define an exact volume necessary for tissue differentiation - it will depend on the characteristics of the pathological process being investigated. If diffuse and involving the whole organ only a limited volume may need to be interrogated. If focal and definable by conventional B scanning methods the defined lesion can de examined. If focal and not definable by conventional methods it may be necessary to examine a large volume of tissue, possibly the whole organ.

Question by: S.N. Rasmussen
Your diagnostic values, were they sonographic or diagnostic predictive figures?

Reply by: R.A. Lerski
The diagnostic accuracies quoted were the percentage of all cases correctly classified as normal or abnormal

		PATHOLOGY	
		abnormal	normal
COMPUTER	ABNORMAL	a	b
RESULT	NORMAL	c	d

That is the figures were $\frac{a + d}{a+b+c+d}$

Question by: D. Nicholas
Can you comment on the overall ability of your analysis to classify all of the tissue groups at once?
What effect has the digitization variation with depth (due to tissue attenuation) on the features you are using?

Reply by: R.A. Lerski
The classification accuracy tends to fall when an attempt is made to design a discriminant function to classify all groups at once. Typically an overall accuracy of 70-75% may be achieved for four tissue groups.

We have carried out a software correction for attenuation on the recorded data - this has little overall effect on the texture analysis results.

Question by: L.F. v.d. Wal
Could you please comment on the surprising fact that your results show high significance in two frequency bands, which show a very low signal to noise ratio, i.e. frequency bands which are positioned outside the bandwidth of your transducer.

Reply by: R.A. Lerski
While it is true that the S/N ratio may be low for some of the low or high frequency bands it is still found that in 117 liver cases statistical significance exists for the normal/abnormal discrimination between these bands. This result would suggest that the experiment could be repeated with a wideband transducer and perhaps with more patients.

Question by: J.M. Thijssen
Don't you have problems when comparing fractions of the spectra, because of the scalloping, when you are performing frequency analysis?

Reply by: R.A. Lerski
The frequency spectra are averaged from the results of 21 A scans (radio-frequency) so that the effects of the scalloping are reduced.

Question by: J.M. Thijssen
What is the exact distance between two (adjacent) points used in the spatial dependence matrix statistics, should the points in a way be statistically independent?

Reply by R.A. Lerski
The points are separated by 10 sample intervals at 20 MHz per sample. This is equivalent to 0.4 mm in tissue and is therefore approximately equal to the wavelength at the centre frequency used of 2.5 MHz. This will tend to reduce the influence of the RF frequency.
Overall the reasons for the usefulness of such techniques are not apparent.

Further discussions and remarks

The compliment was made that the group had presented nice clinical results while using a rather simple signal analysis technique. One reason is that a lot of statistics was employed to arrive at the results, another that all the data were obtained with a single piece of equipment, while keeping the control settings fixed, or taking them into account. The data were taken from patients with diffuse liver diseases, so large tissue volumes could be involved in the analysis. It should be worthwhile to extend the analysis to other groups in order to investigate the general applicability of the results and to find an explanation of some not obviously explained findings, like the significance of data with low signal to noise ratios at low and also at high frequencies. Another approach would be to exchange the data and to let other groups analyse the same data with different mathematical techniques.
A last remark was made, that it seems to become evident that more and more groups are going into A mode video analysis instead of B mode.

ULTRASONIC TISSUE DIFFERENTIATION IN OPHTHALMOLOGY

J. M. Thijssen, M.J.T.M. Cloostermans and W.M. Verhoef
Biophysics Laboratory of the Department of Ophthalmology, University of Nijmegen, Nijmegen, The Netherlands

1. Introduction

Echography is a firmly established method of diagnosis in ophthalmology. In general both the A mode and the B mode examination techniques are applied in a well balanced integration. Since, the A mode displays the reflectivity level of the tissue under investigation most adequately the quantitative echography is performed while using this display mode. The study of kinetic aspects of the pathology is also often performed with the A mode. So the mobility, motility and (neo-) vascularity of lesions is qualitatively assessed. The morphological appearance of lesions is quite conveniently and adequately revealed by a B mode display. The equipment used is mostly a hand held real time mechanical scanner. The intraocular pathology (vitreous) and orbital and periorbital lesions are preferably investigated by B scanning. The quantitative echography is performed in a systematic and, as compared to other disciplines, may be even unique manner. The basic requirement is a standardized and calibrated equipment (A mode). When using this equipment (Kretztechnik: 7200 MA, Biophysics Medical: Triscan) the following parameters are estimated : the reflectivity level, the attenuation (angle kappa, cf. Ossoinig, 1975), the texture, i.e. regularity of the echopattern from solid tissue.
With the clinical experience and knowledge in mind, work at the Biophysics Laboratory has been devoted to the improvement of the clinical diagnosis and generalization of diagnostic data by computer support, to the development of special equipment and to the estimation of acoustical tissue signature parameters.
In this contribution the lines of development are sketched and the results of recent research work are presented from which the outlining of research plans is extrapolated.

2. Instrumentation

The development of equipment started at the Biophysics Laboratory in 1968. At that time no ophthalmological B mode equipment was commercially available. The first clinical results were obtained with a simple linear immersion scanner, while employing Krautkrämer A mode equipment and a Tektronix oscilloscope. The picture shown in Fig. 1 was at that time considered sub-optimal because of the bad axial resolution. The equipment was developed further to include sector and arc scanning modes, as well as linear-sector compound scanning. The electronics was made more sophisticated by adding video differentiation and pulse shaping. The pictures were displayed on a variable persistance oscilloscope and were essentially suitable for optimal display of the anatomy and of the morphology of lesions (Thijssen, 1971, 1972, 1975, 1976).

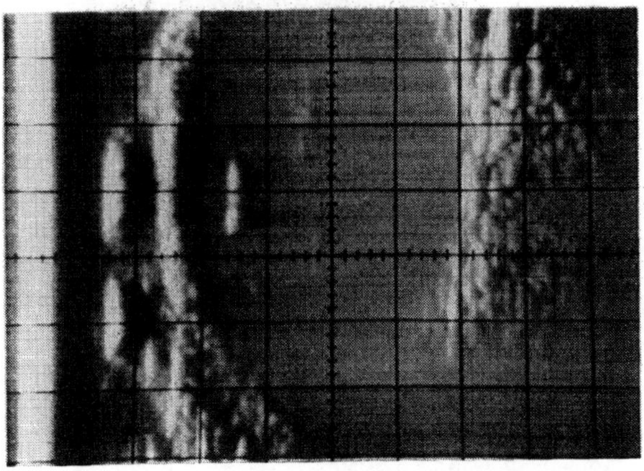

Figure 1. B-mode echogram of the human eye obtained with a simple prototype linear immersion scanner, open shutter technique (Thijssen, unpublished 1969 viz. Thijssen, 1972).

The disadvantages of the immersion technique, both in the patient management and in the accessibility of eye and orbit (interference by eyelids) became apparent. Moreover, the plans for starting the com-

puter analysis of echographic data led to the conviction that a hand held scanner with a real time selection from the B image of an A mode trace was required.
The development of a so-called SAB scanner(Simultaneous A and B mode display) was initiated in 1975 and the design criteria were presented at the 1st Meeting of the World Federation at San Francisco, 1976 (Thijssen et al., 1977). Further optimization and construction resulted in the prototype shown at the 8th Symposium of the International Society for Ultrasound in Ophthalmology (Thijssen, 1981) at Nijmegen, September 1980. The equipment contains a unique scanner head which produces 50 images per second (Fig. 2) and the B mode picture is shown together with a selected A mode trace on a standard TV monitor.

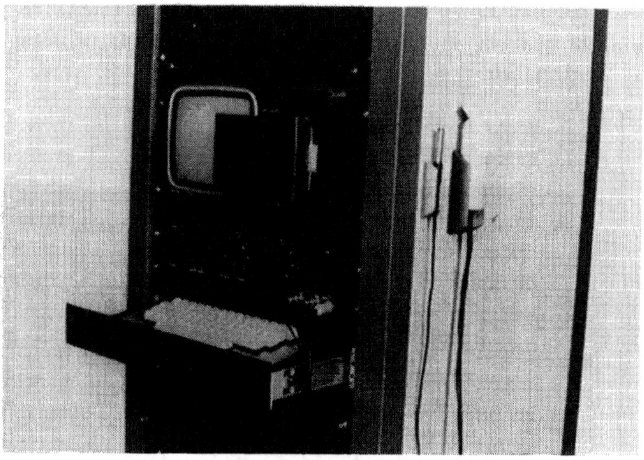

Figure 2. Prototype of the SAB scanner for ophthalmological diagnosis. The viewing monitor is shown left, a polaroid camera is facing a second TV monitor. The scanner head (50 frames per second, TV synchronized) is shown to the right, together with single A probe (Thijssen, 1981).

Further characteristics of the display are (see Fig. 3) : a 16 level gray wedge, a millimeter ruler and two lines of alphanumeric characters. The biometry can easily be performed either during scanning, or after freezing the whole picture in a digital scan converter, by two cursors which cause a video inversion in the A mode display (cf. Fig. 3). The

read-out is found in the right part of the text lines, both in millimeters and in microseconds. The selected A mode trace is available in RF for further analysis by a computer system. The commercial manufacturing of the handpiece for small parts scanning is undertaken by Oldelft Inc. This firm also took part in the construction of the above described prototype of the SAB scanner.
A parallel development was the construction of what was called the electronic tissue model (Van Kervel and Thijssen, 1980; Thijssen and Van Kervel, 1981).

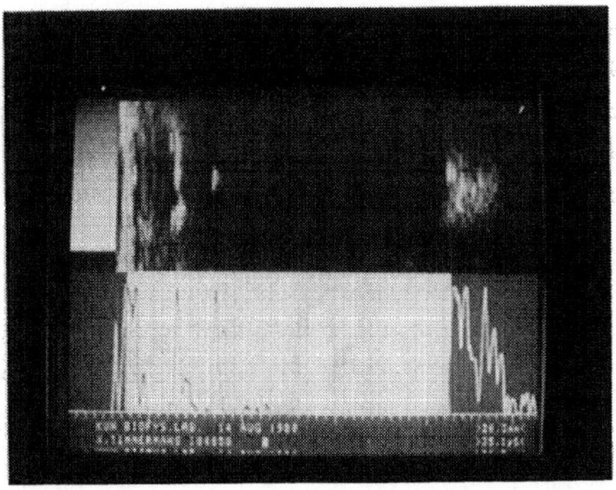

Figure 3. Example of the display of the SAB scanner of a normal eye. Upper part: B-mode, central application on cornea. Lower part: A-mode from central trace in B-mode. Video inversion indicates biometry region (region of interest), length of region indicated in figures bottom right. Ruler in microseconds (Thijssen, 1981).

3. Analysis of video echograms

3.1. A mode video

The quantitative analysis of clinical A mode echograms was undertaken to improve the exactness of quantitative echography and to expand the applicability of clinical knowledge gained with a particular equipment to other equipment. Both goals were served by using the Electronic Tissue model (ETM, see previous chapter) to generate the system gain characteristic curve and storing this curve with the patient's echograms in the computer. It is thus possible to correct the data for non-linear gain compression and to perform the analysis with a true decibel scale (cf. Thijssen et al., 1979a, b, 1981a). Further equipment used is a transient recorder (Biomation 8100) and a digital computer (Digital Equipment Corp. PDP 11/34). The proper analysis is performed after collection of 24 echograms in a 3 seconds acquisition period. The philosophy being that during this period slight movements of the transducer relative to the lesion will occur. The sound beam will "scan" for this reason a small tissue volume, and the 24 echograms will be more or less statistically independent. The analysis is, therefore, based on a tissue volume with enhanced statistical reliability of the estimated quantitative parameters. The output of the computer program is shown in Fig. 4, data are obtained from a proven melanoma of the choroid.

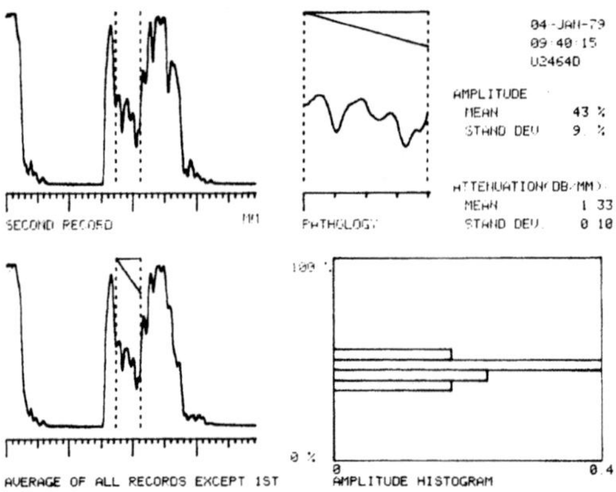

Figure 4. Output of program from statistical evaluation of A-mode video echograms. Left : echograms, upper trace single echogram, lower trace average of 24 echograms. Region of interest indicated by dotted lines.
Top right picture: expanded region of interest, oblique line indicates regression of echopeaks (after decompression of video gain characteristic).
Bottom right picture: 90° rotated amplitude histogram of echopeaks in region of interest (Thijssen et al. 1981a).

The results of a pilot study on 20 cases (30 recordings) of histologically verified melanoma proved the usefulness of the analysis technique (cf. Thijssen and Verbeek, 1981). The differentiation of melanoma from subretinal haemorrhage remains to be investigated, but will be very difficult in case of small lesions because a 3 mm prominence of the lesion is needed at least. The first results on orbital lesions are rather promising, although the detection is often the major problem for the examiner, which is a problem prior to differentiation of course.

3.2. B mode video

The work so far may be characterized as a pilot study to assess the application of image analysis (texture) and image processing techniques to B mode pictures. The images are stored in 6 bits in a 256 x 256 pixel digital memory (Grinell GMR 27 video digitizer) which is interfaced to a PDP 11/34 computer system.
The processing of the images may be divided into several categories:
Preprocessing: the removal of artifacts and limitations of the echographic equipment and transfer into the memory. Examples are: gray scale (histogram) expansion, homomorphic filtering and definition of region of interest.
Point (pixel) processing: this category comprises various kinds of gray scale manipulations like non-linear gray scale coding, windowing, equalizing, etc. The ultimate goal being optimization of the contrasts for the direct visual evaluation.
Picture processing: the information is linearly, or non-linearly filtered in the two dimensional space, or spatial frequency domains. Techniques devised are low pass filtering, high pass filtering, correlation, smoothing and inverse filtering. The goal of these techniques is improvement of resolution and apparent sharpness of the images.
Examples of the application of the processing techniques are shown in Fig. 5 to 7.

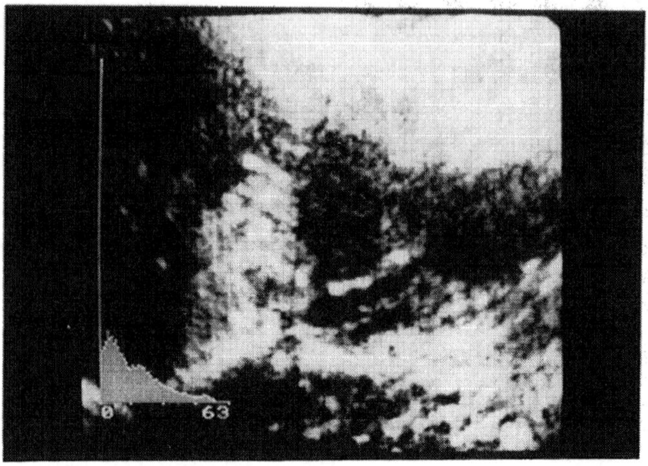

Figure 5. B-mode echogram of liver with metastasis, longitudinal scan. Gray scale histogram of original data (64 gray levels).

Figure 6. Same picture after stretching the original gray levels from 0-31 to the full 64 gray scale.

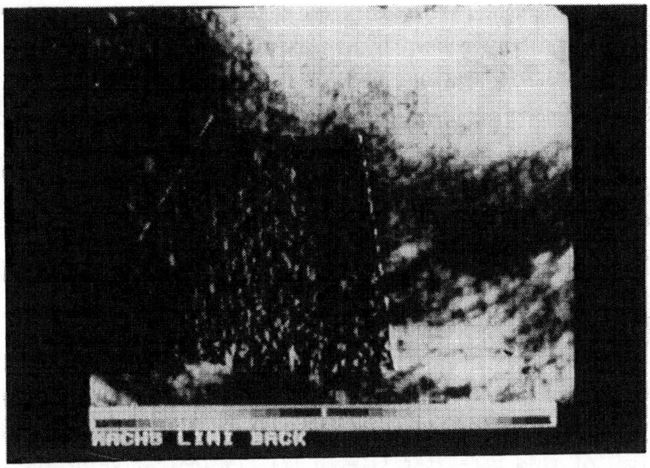

Figure 7. Results of high pass filtering in the region of interest.

The analysis of the images is still in an elementary phase. After the above processing steps statistical analysis may concern the gray scale, the spatial statistics, or both. Examples of the gray scale statistics are the histogram in Fig. 5 and the analysis discussed in chapter 3.1. Spatial statistics could be applied with care, since the axial and lateral resolution of the pictures is essentially asymmetric (cf. Flax et al. 1981), but the techniques described by Haralick et al. (1973) will be seriously considered.

4. Analysis of RF echograms

4.1. Membrane and interface characterization

The differentiation of intraocular membranes from a detachment of the retina, or of the choroid is important clinically.
A means of improving on this is to estimate the exact thickness of the structure and the sign (phase) of impedance discontinuity at the front and back surfaces. We have devised a phase coherent averaging technique which improves the signal to noise ratio of the echograms prior to inverse filtering. The various filtering techniques employed are Hanning-inverse filtering, Wiener inverse filtering and band extrapolation inverse filtering (cf. Bayer and Thijssen, 1981, 1982). Wiener inverse

filtering is chosen, with additional spectral smoothing and time windowing. The improvement of the axial resolution is of the order of a factor two, but it is limited by the transient recorder used for the AD-conversion and intermediate storage before the computer acquisition of the echograms. The devised techniques may be relevant in tumour characterization, since the phase shift at the transition from healthy tissue to a focal lesion may be of diagnostic value.

4.2. Acoustical tissue signature parameters

The parameter primarily being studied is the attenuation coefficient. It is shown in many publications that the frequency dependence (dispersion) of this parameter (i.e. $\alpha_o(f)$) is indicative of the kind of tissue and of the diseased state of some tissues (tumour diagnosis). The in vitro experiments have been dedicated so far to the investigation of the limitations and the feasibility of the estimation of α from backscattered ultrasound energy. The results obtained are rather promising (see Fig. 8 and cf. Thijssen et al., 1981b; Cloostermans and Thijssen, 1982), and the research plans are to extend the scope of the in vitro experiments to different types of healthy tissues (the human eye and abdominal organs) and to biopsy material from the pathology department (diffuse and focal lesions).

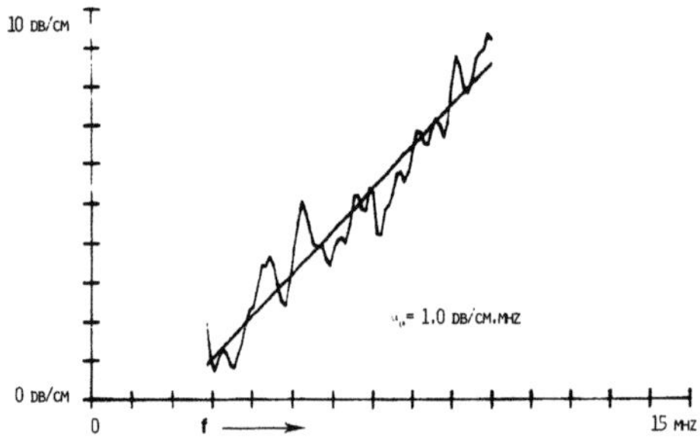

Figure 8. Slope of attenuation coefficient α_o follows from regression line (straight line) through data. Liver tissue preserved in 5% formalin (Thijssen et al. 1981b).

Parallel to the experimental work the theoretical background, i.e. the physics of ultrasound production (Van Kervel and Thijssen, 1982) and propagation, and the modelling of tissues will be undertaken. Furthermore, the construction of a system for applying the devised techniques to in vivo studies in patients (pilot study and clinical evaluation) will be carried out.

Acknowledgements: this work has been supported by the Health Organization TNO, and presently by the Queen Wilhelmina Cancer Foundation.

References

Bayer, A.L. and Thijssen, J.M. (1981): Characterization of membranes by inverse filtering. In: "Ultrasonography in Ophthalmology",pp. 435-442. Eds. J.M. Thijssen and A.M. Verbeek. Junk, The Hague.

Bayer, A.L. and Thijssen, J.M. (1982): Ultrasonic characterization of tissue interfaces by inverse filtering procedures. Subm. Ultrasound Med. Biol.

Cloostermans, M.J.T.M. and Thijssen, J.M. (1982): A beam corrected estimation of the frequency attenuation of biological tissues from backscattered ultrasound. Subm. Ultras. Imag.

Flax, S.W., Glover, G.H. and Pelc, N.J. (1981): Textural variations in B mode ultrasonography: a stochastic model. Ultrason. Imag. 3, 235-257.

Haralick, R.M., Shanmagan, K and Dinstein, I. (1973): Textural features for image classification. IEEE Trans. SMC 3, 610-621.

Kervel, S.J.H. and Thijssen, J.M. (1980): A device for the display and adjustment of non-linear gain curve of ultrasonic A mode equipment. Ultrasonics 19, 40-41.

Kervel, S.J.H. van and Thijssen, J.M. (1982): A calculation scheme for the design of optimum ultrasonic transducers. Subm. Ultrasonics.

Ossoinig, K.C. (1974): Quantitative echography, the basis of tissue differentiation. J. Clin. Ultras. 2, 33-40.

Thijssen, J.M. (1971): A new B-scan system for ophthalmic diagnosis. Ophthal. Res. 2, 285-294.

Thijssen, J.M. (1972): Ultrasonic diagnostics in ophthalmology. Proceedings of 163rd Meeting of the Netherlands Ophthalmological Society, 1969. Ophthalmologica (Basel) 165, 517-518.

Thijssen, J.M. and Gommers, P.A.M. (1975): Methodical and clinical aspects of echo-oculography. Bibl. Ophthal. (Basel) 83, 25-31.

Thijssen, J.M. and Bakker, J.H. (1976): Intensity modulation in B-scan systems. Exc.Med.Internat.Congress Series 363, 115-120.

Thijssen, J.M., Verhagen, C., Kruizinga, R. and Koomen, G.J. (1977): Design criteria for an ophthalmological B-scan system. Pp. 901-908. In: "Ultrasound in Medicine", Vol. 3A. Eds. D.N. White and R.E. Brown. Plenum Press, New York.

Thijssen, J.M., Kruizinga, R. and Dooren, H.A.F.Q. van (1979a): Computer assisted echographic diagnosis. Pp. 12-14. In: "Diagnostica Ultrasonica in Ophthalmologica". Ed. H. Gernet. Remy Verlag, Münster.

Thijssen, J.M., Koomen, G.J. and Verbeek, A.M. (1979b): Computer support for ophthalmic diagnosis. Doc. Ophthal. 48, 315-318.

Thijssen, J.M. (1981): Functional realization of a SAB scanner. Pp. 515-520. In: "Ultrasonography in Ophthalmology". Eds. J.M. Thijssen and A.M. Verbeek. Junk, The Hague.

Thijssen, J.M. and Verbeek, A.M. (1981): Computer analysis of A mode echograms from choroidal melanoma. Pp. 123-130. In: "Ultrasonography in Ophthalmology". Eds. J.M. Thijssen and A.M. Verbeek. Junk, The Hague.

Thijssen, J.M., Bayer, A.L. and Cloostermans, M.J.T.M. (1981a): Computer assisted echo-ophthalmography. Statistical analysis of A mode echograms obtained by tissue sampling. Med. Biol. Engng. & Comp. 19, 437-442.

Thijssen, J.M., Cloostermans, M.J.T.M. and Bayer, A.L. (1981b): Measurement of ultrasound attenuation in tissues from scattered reflections. Pp. 431-439. In: "Ultrasonography in Ophthalmology". Eds. J.M. Thijssen and A.M. Verbeek. Junk, The Hague.

Thijssen, J.M., Bayer, A.L. and Cloostermans, M.J.T.M. (1982): Some results on ultrasonic tissue characterization. In press: "Ultrasound Interaction in Biology and Medicine". Eds. R. Millner and E. Rosenfeld. Univ. Publishing Office, Halle (G.D.R.).

Till, P. (1976): Solid tissue model for the standardization of the echo-ophthalmograph 7200 MA (Kretztechnik). Doc. Ophthalmol. 46, 205-240.

Discussion

J.M. THIJSSEN, M. CLOOSTERMANS and W.A. VERHOEF

Acoustospectrography of neoplasms

Question by: G. Berger
In order to make a correction of the distance influence on the spectra don't you think that you have to take into account the evolution of the spectra measured on a diffracting particle rather than a flat reflector?

Reply by: M. Cloostermans
For a correct estimation of the correction of the distance influence an integral over the beamwidth of the power spectra obtained from point reflectors should be measured and evaluated. This is experimentally not possible at this moment, because a great accuracy of the manipulators is not available at this moment. Therefore we simulated this integral with a flat plate reflection.

Question by: K-M. Irion
What happened in case of inverse filtering if there are frequency shifts in the echo's spectra between two layers (membranes).

Reply by: M. Cloostermans
If the membrane has a non-constant transfer function the result of inverse filtering is not correct. The second pulse will be a convolution of the transfer function of the membrane and the ideal pulse obtained from a membrane which has a constant transfer function.

Further discussions and remarks

The influence of the sound field characteristics of a focused transducer on the spectral contents of the echogram from a flat reflector, as well as on the estimation of the attentuation coefficient is relatively high. In solid tissue the effect has been shown to be less,

which fits in the known defocusing effect due to scattering.
The ophthalmological scanner will not become available in a commercial way, but the Oldelft firm which cooperated in the project produces the patented handpiece now in three different versions, linear, sector and trapezoid scanning methods, in OEM for other firms.

SURVEY OF ULTRASOUND WORK AT ISPRA

I. Bredael
Commission of the European Commuinities, Joint Research Center of ISPRA - Italy
Material Science Division, Non Destructive Testing Laboratories

The present activities can be grouped mainly in the characterization of instruments and of materials.

1. Characterization of instruments

- single, dual, annular and array probes.
- analog, digital, optical equipment using the following methods:
  - T/R water tank with ball, plane block or phantoms
  - Schlieren bench beam analysis (longitudinal)
  - Liquid crystals beam analysis (continuous waves)
  - Laser interferometer (displacement $> 0{,}01 A^{o}$)
  - Electrical measurements (Z, ϕ, efficiency)

All these tests can be conducted considering the unknown instrument as a black-box (no internal acces for signal injection or electrical output).

2. Characterization of materials

- pulse or burst technique (monochromatic or chirp)
- time or frequency domain signal acquisition as analog or digital information
- 1 D, 2 D or 3 D processing and analysis
- false gray or false color display code
- system feasibility studies
  - system analysis
  - system specification
  - system simulation

  for existing or future equipment

Some specific projects that have been running since 1976

- Collaboration with the Brugmann Hospital - Brussels (Dr. Lévi) for the biological data digitalization and analysis in utero cancer examinations
- Artificial eye testing block for ophthalmological instruments calibration and testing (University of Ferrara - Italy)
- Interaction studies between human blood red cells and ultrasound at diagnostic levels (Prof. Pinamonti-Ferrara-Italy). Absolute acoustic intensity measurements
- Instrumentation characterization (Organon-Technica, Smith & Kline, Gilardoni, Siemens, Philips, Kretz, OTE-Biomedica, etc.)
- Medical probes characterization (more than hundred types)
- Material evaluation (absorption, dispersion, etc.)
- Image analysis (cardiology, obstetrics, internal medicine, X-rays, RZ)
- Actual work : Tissue modelling in analogy with hard materials considering the plane multiple interface model and the grain or core model. These models include the absorption, resonance, diffraction, scattering and diffusion parameters.

GPSR Compliance

The European Union's (EU) General Product Safety Regulation (GPSR) is a set of rules that requires consumer products to be safe and our obligations to ensure this.

If you have any concerns about our products, you can contact us on ProductSafety@springernature.com

In case Publisher is established outside the EU, the EU authorized representative is:

Springer Nature Customer Service Center GmbH
Europaplatz 3
69115 Heidelberg, Germany

Batch number: 10607224

Printed by Printforce, the Netherlands